Threats to Food & Chain Infrastructure

MW01613939

Edited by

Virginia Koukouliou

Greek Atomic Energy Commission
Department of Environmental Radioactivity
Agia Paraskevi, Greece

Magdalena Ujevic

Croatian National Institute of Public Health
Zagreb, Croatia

Otto Premstaller

Federal Ministry of Agriculture
Forestry, Environment and Water Management
Vienna, Austria

Springer

Published in Cooperation with NATO Public Diplomacy Division

Proceedings of the NATO Advanced Research Workshop on Threats to Food and Water
Chain Infrastructure
Vienna, Austria
3–5 December 2008

Library of Congress Control Number: 2009942083

ISBN 978-90-481-3545-5 (PB)
ISBN 978-90-481-3544-8 (HB)
ISBN 978-90-481-3546-2 (e-book)

Published by Springer,
P.O. Box 17, 3300 AA Dordrecht, The Netherlands.

www.springer.com

Printed on acid-free paper

Threats to Food and Water Chain Infrastructure

NATO Science for Peace and Security Series

This Series presents the results of scientific meetings supported under the NATO Programme: Science for Peace and Security (SPS).

The NATO SPS Programme supports meetings in the following Key Priority areas: (1) Defence Against Terrorism; (2) Countering other Threats to Security and (3) NATO, Partner and Mediterranean Dialogue Country Priorities. The types of meeting supported are generally "Advanced Study Institutes" and "Advanced Research Workshops". The NATO SPS Series collects together the results of these meetings. The meetings are coorganized by scientists from NATO countries and scientists from NATO's "Partner" or "Mediterranean Dialogue" countries. The observations and recommendations made at the meetings, as well as the contents of the volumes in the Series, reflect those of participants and contributors only; they should not necessarily be regarded as reflecting NATO views or policy.

Advanced Study Institutes (ASI) are high-level tutorial courses intended to convey the latest developments in a subject to an advanced-level audience

Advanced Research Workshops (ARW) are expert meetings where an intense but informal exchange of views at the frontiers of a subject aims at identifying directions for future action

Following a transformation of the programme in 2006 the Series has been re-named and re-organised. Recent volumes on topics not related to security, which result from meetings supported under the programme earlier, may be found in the NATO Science Series.

The Series is published by IOS Press, Amsterdam, and Springer, Dordrecht, in conjunction with the NATO Public Diplomacy Division.

Sub-Series

A.	Chemistry and Biology	Springer
B.	Physics and Biophysics	Springer
C.	Environmental Security	Springer
D.	Information and Communication Security	IOS Press
E.	Human and Societal Dynamics	IOS Press

http://www.nato.int/science
http://www.springer.com
http://www.iospress.nl

Series C: Environmental Security

PREFACE

The malicious contamination of food and water for terrorist purposes is a real and current threat. Deliberate contamination of food and water at a single location could have global public health implications. The workshop responded to increasing concern in Member States that chemical, biological or radionuclear agents might be used deliberately to harm civilian populations and that food might be used as a vehicle for disseminating such agents. The food supply chain has seen an increasing trend in globalization of food sourcing and thus increasingly complex supply chains. Increasing globalization of food trade means that an inability to respond to a food emergency could have significant consequences on health and trade in many countries (World Health Organization 2002). Governments also have a role in facilitating preventative food safety through both voluntary and regulatory mechanisms (World Health Organization 1996). Food terrorism has been defined (World Health Organization 2002) as:

... an act or threat of deliberate contamination of food for human consumption with chemical, biological or radionuclear agents for the purpose of causing death to civilian populations and/or disrupting social, economic or political stability.

During the NATO Advanced Research Workshop, a detailed review of the characteristics of biological, chemical and radioactive agents that would make them attractive for use as possible weapons against a given population, as well as the potential consequence bio- or agro-terrorism attack were presented by the speakers, often using actual examples. There have been many instances where civilian food supplies have been sabotaged throughout recorded history, during military campaigns and, more recently, to terrorize or otherwise intimidate civilian populations. Deliberate contamination of food by chemical, biological or radionuclear agents can occur at any point along the food chain, from farm to table, the vulnerability depending on both the food and the agent. For example, in 1984, members of a religious cult contaminated salad bars in the USA with *Salmonella typhimurium*, causing 751 cases of salmonellosis, and more recently on spring of 2007 a massive intentional contamination of wheat gluten used as an ingredient in pet food with a fertilizer called melamine and in fall of 2008, a similar concern with melamine surfaced again this time in human food.

The deliberate act to contaminate food or water supplies with radioactive materials is unlikely to lead to the significant internal contamination of a large number of people due to the enormous quantities of radioactive material that would be required to reach high levels of contamination in mass-produced or distributed supplies. Although, based on data presented at the Workshop concerning the more than 30,000 missing radioactive sources all over the word, the radioactive contamination of food or water is also a scenario that must be taken seriously into consideration.

During the last two decades there have been several emerging hazards linked to animal diseases or originating in animal products for example: Avian Influenza (AI), Bovine Spongiform Encephalopathy (BSE), West Nile Fever, Severe Acute Respiratory Syndrome (SARS), and Ebola virus. All these diseases or events directly or indirectly affect food security and/or food safety. Approximately 75% of all emerging diseases are zoonotic by either an association with animal populations or an evolution of the disease in animals making it possible to move from animal species to humans.

Participants were presented the primary results of the ongoing NATO-SPS Pilot Study on "Food Chain Security". These results focused mainly on (i) an overview of the food system; (ii) prevention, surveillance and detection systems and (iii) response system. The importance of issues such as: vulnerability assessments, risk communication in risk analysis, risk perception, traceability, preparedness – awareness, communication, have to be considered when working on food chain security.

The food and water supply utilities have been starting to implement surveillance and preparation of guidance papers. The most authoritative example is the 2002 World Health Organization document "Terrorist Threats to Food". The WHO position is that counterterrorism should be considered as integrated into the wider programmes for food safety, including among others increased surveillance and rapid response to outbreaks. Current crises in the area of food safety can be aggravated by the transboundary interconnection between industries and retailers. In Europe, the spreading of disease from farm animals has highlighted the unexpected vulnerability of our meat supply system, amplified also by the media coverage and by the absence or the delay in redress measures. It is necessary to coordinate and harmonise prevention, restoration and redress, utilising also the experience recently obtained within the framework of the NATO Science for Peace project "Development of a prototype for the International Situational Centre on Interaction in Case of Ecoterrorism".

Consumer trust in food, and food safety systems has been on the agenda over the last 10–15 years, brought forth by major events linked to food safety and quality as well as by structural and political changes in the food system. Research has shown that consumer responses must not be influenced by unbalanced media presentations ('scares') or personal uncertainty in view of technological innovations (like GM food). Such aspects are certainly important, but the key factor triggering responses of distrust seems to be how market and public actors handle food safety issues. Lack of accountability and transparency and disregard of consumer expectations and interests form important explanations to negative reactions among consumers. Media presentations are important in communicating such problems, which of course also may be exaggerated by the media. But lack of openness and responsiveness on behalf of governments seems to be more of a problem than media amplification of the stories. Historically, it is in periods of rapid change that we find most cases of consumer mobilisation and food riots. But responses among consumers and the public do not necessarily depend on the emergence of new conditions or problems per se, but on whether these challenges are met in ways that consider consumer concerns and interests.

Dependence on imported food or non-locally produced food, which is the result of both the globalisation and changing of consumer mentality, also poses a threat for food security in the case of terrorist actions, natural disasters or all kinds of emergency situations. Food coming from outside of a given country may not have the same standard of food safety applied to the growing and processing on the food. Imported foods could be contaminated, contain harmful substances etc., they may not be fresh, as long distance transport requires them to be chemically treated. Also the transportation could pose a big problem, resulting sometimes even dangerous or impossible situations due to the lack of fuel or vehicles or destroyed transportation ways. For assuring food security in general and in crisis situations it is not sufficient to simply establish adequate food reserves, to develop new more productive varieties and technologies, or to have plans for storage and distribution of food reserves, it is of paramount importance to assure more sustainable food supply with the promotion and support of local production and processing of food.

A study concerning deliberate threats against food chain has been carried out in France, using two methods: analysis of feasibility and impact on public health; data collections of former biological and chemical criminal or malevolent outbreaks. Twenty five biological agents have been classified in three increasing danger categories. The same approach has been performed for 71 chemical agents. It should be noticed that many chemical agents are not affected by temperatures of foodstuffs treatments and that some chemical agents are easily available on the market or purchased on the WEB. In food plants the implementation of Hazard Analysis Control Critical Points (HACCP) method aims at controlling sanitary dangers and avoiding accidental outbreaks. This kind of method can also be developed to prevent criminal intentions. Each food plant can implement his specific program concerning prevention management of threats, securing access to equipment, premises and products, and to control people movements.

The Final report on "Vulnerability of water supplies of irrigation, Livestock and Food processing" provides a description on how to identify the weak points of the supply chain of water for irrigation, livestock and food processing and was noted by the FAPC Plenary on March 2005. Documents such the ISO TC 164 Water Supply WG 15 "Security of drinking water supply", can help the countries to enhance security, citizens safety and crisis management capacity in an increasingly interdependent and borderless world. Examples of comprehensive and substantial water policy activities could be drawn by counties witch implement water crisis management systems and one important aspect that has to be addressed is the permanent practical training and the periodical check and review of emergency operations at a larger scale. There is the clear need to deal with the issue of emergency water supply within exercises and practical training at all levels and in all dimensions. One of the most important aims of training should be to raise awareness of water suppliers for emergency situations and to improve the co-operation of water supply operators with emergency organisations.

In a system as complex as the food continuum, it is almost impossible to tackle all hazards related to potential intentional contamination. The challenge for those who are engaged in food defence is to narrow down the list of potential threat agents based on their characteristics in an attempt to focus resources and efforts on the most likely agents and food matrix combinations. The attributes that make food and water very attractive as a target can also make it a very challenging target.

Vienna, Austria   V. Koukouliou

CONTENTS

Session II   Management

CONTRIBUTORS

Alpas, Hami
Department of Food Engineering, Middle East Technical University,
Ankara 06531, Turkey
imah@metu.edu.tr

Astrid Rompolt
Stadt Wien – Wasserkerke, Grabnergasse 4-6, 1060 Vienna, Austria
astrid.rompolt@wien.gv.at

Czerni, Wolfgang
Geschäftsführer der Infraprotect GmbH, Gesellschaft für Risikoanalyse,
Notfall- und Krisenmanagement, Wolfsaug. 11/6/13, A-1200 Wien
w.czerni@infraprotect.at

Dadic, Željko
Croatian National Institute of Public Health, Rockefeller st. 7,
Zagreb, Croatia
zeljko.dadic@hzjz.hr

Hansen, Chris
1897 Beach Road, RR#2, Oxford Mills, Ontario KOG 1SO, Canada
Chris.Hansen@inspection.qc.ca

Kjaernes, Unni
The National Institute for Consumer Research, PO Box 4682,
Nydalen N-0405 Oslo, Norway
unni.kjarnes@sifo.no

Koç, Ahmet Ali
Department of Economics, Akdeniz University, Antalya, Turkey

Koukouliou, Virginia
International Atomic Energy Agency, NSRW-Technical Assistance and
Information Management Unit, Regulatory Infrastructure and Transport
Safety Section, Division of Radiation, Transport and Waste Safety,
Department of Nuclear Safety and Security, Wagramerstrasse 5.
PO Box 100, A-1400, Vienna, Austria
V.Koukouliou@iaea.org

Madelaine Maillot, Evelyne
Veterinary Public Health, General Counsel for Agriculture,
Food and Rural Areas, Ministry of Agriculture and Fisheries, France
evelyne.maillot@agriculture.gouv.fr

Maestri, Elena
Department of Environmental Sciences, University of Parma,
Viale Usberti 11/A, 4310 Parma, Italy

Marmiroli, Nelson
Department of Environmental Sciences, University of Parma, Viale
Usberti 11/A, 4310 Parma, Italy
nelson.marmiroli@unipr.it

Marmiroli, Marta
Department of Environmental Sciences, University of Parma,
Viale Usberti 11/A, 4310 Parma, Italy

Premstaller, Otto
Deputy Head of Division III 3, Federal Ministry of Agriculture,
Forestry and Water Management, Stubenring 1, A-1012 Vienna, Austria

Rompolt, Astrid
Stadt Wien – Wasserwerke, Grabnergasse 4-6, 1060 Vienna, Austria
astrid.rompolt@wien.gv.at

Santamato, Stefano
NATO, Civil Emergency Planning, Operations Division,
NATO HQ – Brussels
santamato.stefano@hq.nato.int

Schimon, Wilfried
Ministry for Agriculture, Foresty, Environment and Watermanagement,
Marxergasse 2, 1030 Wien, Austria
wilfried.schimon@lebensministerium.at

Aşci, Serhat
Department of Economics, Akdeniz University, Antalya, Turkey

Ujevic, Magdalena
Croatian National Institute of Public Health, Rockefellerova 7,
10000 Zagreb, Croatia
magdalena.ujevic@hzjz.hr

Urbancic, Alenka
Ministry of Education and Sport Masarykova 16, SI - 1000 Ljubljana,
Slovenia
alenka.urbancic1@gov.si

Vitale, Ksenija
School of Public Health "A. Stampar", Rockefeller st. 4,
Medical School, University of Zagreb, 10000 Zagreb, Croatia
kvitale@snz.hr

THE FOOD AND AGRICULTURE PLANNING COMMITTEE (FAPC)

NATO's Civil Emergency Planning activities aim to ensure effective use of civil resources for use during emergency situations, in accordance with Alliance objectives. It enables Allies and Partner nations to assist each other in preparing for and dealing with the consequences of crisis, disaster or conflict.

NATO's eight Planning Boards and Committees bring together national government experts, industry experts and military representatives to provide coordinated planning across various sectors of civil activity. Under the guidance of NATO's Senior Civil Emergency Planning Committee, and within the framework of Ministerial Guidance for Civil Emergency Planning, these bodies advise on crisis-related matters and assist NATO Military Authorities and nations to develop and maintain arrangements for effective use of civil resources.

This leaflet provides a brief overview of the Food and Agriculture Planning Committee (FAPC) which was established by the North Atlantic Council on 28 November 1952. It is responsible in peacetime for co-coordinating and monitoring national and NATO arrangements for civil emergency preparedness and crisis management in the areas of food, agriculture and water. The FAPC carries out its tasks within the framework of the overall aims of NATO Civil Emergency Planning.

FAPC Responsibilities

FAPC's remit is to ensure that expertise and experience in the food, agriculture and water sector serves NATO's civil and military needs to maximum effect through the exchange of information and best practices.

Its responsibilities involve:

- The development of training programmes aimed at sustaining the essential continued interest and participation of food, agriculture and water experts in NATO crisis management arrangements and maintaining such arrangements in an adequate state of readiness

- Pursue the joint civil-military planning necessary to provide a NATO civil capability to support food, agriculture and water requirements in a crisis

- Support CEP-related activities with Co-operation Partners

- Monitor, review and exchange information on: the current organisation of the food, agriculture and water sectors; the activities of other international organisations in these sectors; national legislation and arrangements for emergency food, agriculture and water supply; and review FAPC activities with the aim of ensuring economic use of resources available to the Committee and avoiding duplication of the product of other international organisations

Format

The Food and Agriculture Planning Committee, on which all NATO member and Partner nations are entitled to sit, normally meets twice a year in Brussels at NATO HQ. However, much of the FAPC's work is achieved within smaller working Groups which are set up depending on the issues to be examined or events to be prepared.

Vital Resources Seminar

Every 2 years, the FAPC Plenary meeting is hosted by a nation back-to-back with a Vital Resources Seminar which addresses topical issues. The Vital Resources Seminar brings together national representatives and national Military Authorities as well as private sector experts from Allied, Partner and Mediterranean Dialogue nations.

Participants are familiarised with NATO's priorities and challenges and share information on security concerns related to the food, agriculture and water sectors. The Vital Resources Seminar, in responding to the new security challenges, addresses a range of issues such as examining and suggesting possible remedies to the problems of critical infrastructure protection in the food, water and agriculture sectors, its interdependencies and vulnerabilities. The seminars also address the impact of the loss of food, water or agriculture (and the resources needed to recreate these services) on the wider infrastructure and to determine the measures necessary to re-establish adequate and safe production levels. This activity is facilitated via a scenario driven exercise.

Civil Experts

One of the major responsibilities of the Planning Boards and Committees is to maintain a pool of civil experts prepared to provide advice in an evolving crisis and in the planning for military operations. Civil experts provide advice on the use of civil resources to the Council, the SCEPC (in NATO or EAPC format), NATO Military Authorities (NMAs), nations or other appropriate bodies as agreed by the Council/SCEPC. There may also be a need for civil experts to support nations and international organisations in case of large-scale emergencies.

Civil Experts are individuals made available by nations to NATO who are recruited, selected, and trained by the Planning Boards and Committees in accordance with the SCEPC and the Civil Emergency Planning Crisis Management Arrangements.

Civil Experts are nationals of Allied or Partner countries and can be representatives of industry/business or government/administration.

Vulnerabilities and dependencies in an evolving environment

One of the Committee's concerns is the identification of vulnerabilities and dependencies of member nations vis-à-vis the security of national food supplies, the operation of their agri-food sector and the robustness of water supply in an evolving international context.

- Vulnerable natural resources: Resource production activities are effected via the exploitation of natural environments (climates, soils and water) and the use of living beings (animals; micro-organisms; plants).
- Uncontrollable and unchangeable biological mechanisms: The processes that can be used by operators are linked to evolving cyclical systems (seasons; reproduction).
- Perishable foodstuffs: The foodstuffs offered to consumers must be healthy, acceptable and nutritional.
- The Committee takes into account the demands of changing societies:
- Technological change: the specialisation of companies; the sophistication of processes; the mechanisation and automation of activities; reductions in stock levels; rising energy consumption; just-in-time trade; health regulations.
- Sociological change: the urbanisation of consumers; the development of quick-frozen food; higher consumption outside the home; ever-increasing mass population movements.
- Economic change: the concentration of companies; the regionalisation of primary production; community (EU) regulations; international agreements on competition (WTO).

Contrary to certain committees which work in areas that are partially or totally national responsibilities (the Military; Police; Customs; firefighters; telecommunications; solar energy), the FAPC is involved in areas of endeavour which, in normal times, are occupied by private operators, who are essentially subject to economic imperatives, market fluctuations and international trade.

Cooperation

The Committee maintains practical co-operation with the Food Hygiene Technology and Veterinary Services Team of Experts, a working group of the Committee of the chiefs of Military Medical Services (COMEDS) and the Science for Peace and Security Committee in the areas of food security and water management.

FAPC maintains contacts with other international organisations such as the World Organisation for Animal Health (OIE). FAPC is also at the forefront of co-operation with Mediterranean Dialogue countries.

Practical examples of FAPC activities and projects

NATIONAL MEASURES AGAINST THE SPREAD OF AVIAN INFLUENZA

The FAPC acknowledges that primary responsibility in dealing with the possible spread of avian influenza remains with national authorities and with international organisations such as the WHO and the OIE. However, given the importance of the subject and the interest that EAPC countries expressed in sharing relevant information and best practices, the FAPC has developed a consolidated report bringing together the various measures that nations are implementing against the spread of avian Influenza.

VIRTUAL LIBRARY

FAPC is working on setting up a virtual library of national best practices. This library will house best practices in the areas of food chain security and vector borne diseases (such as avian influenza). Its aim is to improve national ability to handle crisis situations by the sharing of national best practices and experiences in the agriculture, food and water sectors.

IMPACT OF PESTICIDES ON DEPLOYED TROOPS

The FAPC is working on a study on the possible impact of pesticides and other chemical substances used in the agricultural sector on deployed troops.

SESSION I
THREAT

RADIONUCLEAR MATERIAL AGENTS THAT COULD BE USED IN FOOD AND WATER SUPPLY TERRORISM

VIRGINIA KOUKOULIOU

Department of Environmental Radioactivity, Greek Atomic Energy Agency, Division of Licensing and Inspections, P.O. Box 60092 15310, Agia Paraskevi, Greece
e-mail: vkoukoul@eeae.gr

Abstract: The contamination of food or water supplies with radioactive materials centers the attack on the ingestion pathway, where the aims may be to: expose the public who consume the contaminated food or drink the contaminated water; stop the provision of food or water supplies to the public; and cause widespread panic and public alarm. The radiological consequences may include: contamination of water treatment plants, service reservoirs, header tanks and water supply systems; contamination of food products, wholesale food markets, supermarkets or food processing facilities; and the loss or disruption of the water and/or food supply chain.

The occurrence of immediate fatalities or casualties suffering from the effects of radiation exposure via the ingestion pathway is very unlikely since extremely large amounts of radioactive material would be required to achieve sufficiently high concentrations and, even if this occurs, it is very unlikely that it would affect a large number of people.

The radionuclides that can be used or released during a radiological emergency, where a significant radiation dose could be received as a result of consumption of contaminated food, could be:

- Nuclear reactors (I-131; Cs-134 + Cs-137; Ru-103 + Ru-106)
- Nuclear fuel reprocessing plants (Sr-90; Cs-137; Pu-239 + Am-241)
- Nuclear waste storage facilities (Sr-90; Cs-137; Pu-239 + Am-241)
- Nuclear weapons (i.e., dispersal of nuclear material without nuclear detonation) (Pu-239)
- Radioisotope thermoelectric generators (RTGs) and radioisotope heater units (RHUs) used in space vehicles (Pu-238)

The radionuclides listed above are expected to be the predominant contributors to radiation dose through ingestion in the most of the scenarios. When more than one radionuclide is released, the relative contribution that a radionuclide makes to radiation dose from ingestion of subsequently contaminated

V. Koukouliou et al. (eds.), Threats to Food and Water Chain Infrastructure,
DOI 10.1007/978-90-481-3546-2_1, © Springer Science+Business Media B.V. 2010

food depends on the specifics of the accident and the mode of release. In unique circumstances other radionuclides (like Po-210) may contribute radiation doses through the food ingestion pathway.

Although the deliberate act to contaminate food or water supplies with radioactive materials it is unlikely (though not impossible), there is a need to co-operate with radiological experts and media specialists to quickly assess the potential medical impact of such acts and provide public information to alleviate fears in the potentially affected public. There is also a need to develop a plan, at the national level, to monitor a representative sample of the potentially affected population to confirm the health risk assessment and reassure the public.

Keywords: Dirty bombs, food contamination, radiation detriment, radiation effects, radiation sources

1. Radiation detriment

1.1. THE HUMAN BODY

The human body is affected by ionizing radiation from sources outside the body (an external source) and from radioactive materials which have entered the body (internal contamination).

For external sources to cause harm they must be able to penetrate the body and affect our cells. As alpha particles and low energy beta particles cannot penetrate the skin (unless through an open wound or through chemical absorption), only high energy beta particles, gamma rays, x-rays and neutrons present an external radiation hazard.

If radioactive materials are taken into the body either by inhalation, ingestion, through an open wound or through chemical absorption, they are in direct contact with cells and can therefore irradiate all of the cells surrounding them. In this case, alpha and beta particles are of most concern as they both have short ranges in biological tissue and will dissipate their energy in a small volume. However, when considering sources inside the body, it is important to have some idea of how they human body works and how radionuclides can become deposited in the various parts of the body.

The harm or potential harm resulting from exposure to ionizing radiation is expressed as the equivalent dose (H). The term takes into account:(a) the total energy absorbed per unit mass of a tissue or organ, the absorbed dose (D); and (b) the type and energy of the radiation causing the dose, which influences how the energy is distributed in the tissue, represented by a radiation weighting factor (w_R).

The absorbed dose is the measured number of joules per kilogram (J/kg) expressed as a unit called the gray (Gy).

One joule per kilogram of absorbed dose averaged over a tissue or organ T(rather than at a point) and weighted (multiplied by the appropriate wR) for the radiation quality of interest, expresses the equivalent dose (H) as a unit called the sievert (Sv). The unit of equivalent dose is the joule per kilogram with the name sievert (Sv). When a tissue or organ is exposed to several different types of radiation, the potential harm to it is the sum of the equivalent doses for each radiation quality as indicated by the formula. HT is the sum of the individual equivalent doses and DT, R is the absorbed dose due to radiation R. For example, if an organ receives 1 Gy of gamma radiation (w_R of 1 for photons), its equivalent dose is 1 Sv. A further absorbed dose of 1 Gy due to 4 MeV neutrons (w_R of 10) would add 10 Sv. The organ has received 11 Sv equivalent dose (harm). Absorbed dose was formerly expressed in a unit called the rad (equal to 0.01 Gy) and equivalent dose was formerly called dose equivalent and expressed in a unit called the rem (equal to 0.01 Sv).

One sievert is a very large equivalent dose.

Naturally occurring radioactive materials in the environment and food, in addition to cosmic radiation, typically provide a background dose of a few millisieverts (thousandths of a sievert, mSv) per year to members of the public.

The medical use of x rays may deliver tens or hundreds of microsieverts (millionths of a sievert, μSv) per investigation to a patient. Occupational exposed workers are commonly exposed to dose rates of tens of microsieverts per hour (μSv/h) up to several millisieverts per hour (μSv/h) for brief periods and accumulate a dose equivalent of millisieverts per year as a consequence.

1.2. RADIATION DAMAGE

1.2.1. Cell damage

Radiation damage occurs when ionizing radiation interferes with the normal operation of a cell causing direct and indirect damage to the cell.

Direct damage occurs when radiation strikes a critical area of the cell (such as the DNA) and causes direct ionization in the molecule itself.

Indirect damage occurs as the result of the formation of very chemically reactive atom groups, called free radicals, within the cell.

Radiation damage to individual cells results almost entirely from the damage to the DNA. The three main consequences of this damage can be summarized as follows:

- The cell may die.
- Changes may occur in the cell which might lead to abnormal cell division.
- The genetic material of the cell may change and the change is passed on to new cells.

Note that ionizing radiation does not cause any damage to cells that could not be caused by other agents such as chemicals and viruses.

1.3. EFFECTS OF RADIATION ON HUMANS

Ionizing radiation acts on the cells of the human body. If the cells do not repair themselves, permanent effects of radiation damage can be seen as biological changes in tissues and organs. These changes may be manifested as medical symptoms which are classified into deterministic or stochastic effects. These effects are of particular concern in the case of the developing foetus.

1.3.1. Deterministic effects

The most common result of radiation damage is for the cell to die. If only a few cells are affected, it is not usually a problem as there are many cells in the body and new cells will replace the dead cells. However, as the amount of radiation absorbed (i.e. the dose) increases, a point will be reached where sufficient cells are killed to affect the overall operation of the organ. The result of this is a loss of organ function which will become more serious as the number of affected cells is increased.

The different types of radiation damage resulting from the loss of organ function are known as deterministic effects. These effects are characterized by having a threshold dose (below which there is no observable effect) followed by a response where the severity of the effect increases with increasing radiation dose. An example of a deterministic effect is erythema or skin reddening. Exposure to a low dose of ionizing radiation, below the threshold dose, will not cause the skin to redden. If the dose is increased to a value just above the threshold dose the skin will redden, in the same way that fair skinned people experience mild sunburn. If the dose is increased further, blistering will occur (like severe sunburn) and higher doses will result in death of the skin tissue and ulceration. Other deterministic effects results from irradiation of a particular organ include sterility (temporary or permanent) and cataracts.

Note that the effect of radiation dose on a particular person depends on biological factors (e.g. age and general health) as well as on chemical factors (e.g. the amount of oxygen present in tissue). Therefore in any population there will be a range of sensitivity to radiation. Hence, the threshold dose in a given tissue will be reached at lower doses in more sensitive individuals. As the dose increases more individuals will show the effect, up to a dose above which all exposed people will show the effect.

Deterministic effects are most often seen in cases of high doses of radiation delivered in a short period of time (i.e. in the case of acute exposure). Other than for controlled medical exposure, high doses are not usual in the workplace.

Hence deterministic effects are only seen in accident situations and do not occur routinely in the workplace.

The severity of the deterministic effects mentioned in Table 3 depends on the size of the dose and the period over which the dose was received. In fact, if the dose is received over several weeks rather than all at once, the threshold dose at which the effect occurs increases considerably, usually by about 100%.

Very high doses of radiation to the whole body can cause sufficient damage to the organs to stop their function and this may ultimately cause death. Radiation sickness (nausea, vomiting diarrhoea) is an early deterministic effect resulting from an acute high dose to the whole body.

1.3.2. Stochastic effects

Sometimes, the effect of radiation is not to kill the cell but to alter it in some way. In most cases this alteration will not affect the cell significantly so there will be no observable effect. However there is a possibility that the injury might affect the control system of the cell, subsequently causing it to divide more rapidly than normal. If the affected cell does begin to divide in this way, an increasing number of abnormal daughter cells will be produced. If these abnormal cells invade normal tissue they are called malignant cells and this results in cancer.

The type of cancer formed is dependent on the type of the original cell which was altered. Cancers do not appear immediately after radiation exposure but appear after a latency period in which no effects are observable. The latency period is dependent on the type of cancer but can vary from 2 years for leukemia to 30 years or possibly longer for some solid cancers. Hence, cancer is classified as a late effect.

Unlike deterministic effects, the amount of radiation exposure does not change the severity of the cancer but it does alter the chance of getting cancer. In other words, exposure to a high dose can increase the risk of getting a cancer but if cancer occurs (whether it be at low or high dose) the severity of the cancer is the same. This is rather like a lottery prize in that even a single ticket could win the first prize but the more tickets you buy the more chance you have of winning. However, the value of the prize does not change. Effects of ionizing radiation which rely on chance are referred to as stochastic effects.

For the purposes of radiation protection, it is assumed that the probability of a stochastic effect increases linearly as the dose increases and that there is no threshold dose. If there is no threshold dose then it is considered that even small doses of radiation might cause cancer.

Stochastic effects are the only effects possible at low doses. The risk of stochastic effects is the primary reason for limiting doses to both the public and radiation workers.

Stochastic effects are the only effects possible at low doses and hence, radiation protection is aimed at preventing deterministic effects and reducing the chances of stochastic effects occurring.

1.3.3. Hereditary effects

If one of the reproductive cells (either the sperm or the ovum) is damaged by ionizing radiation, there is a chance that this damage may affect either the immediate child or subsequent generations. This type of effect is therefore known as a hereditary effect. Hereditary effects are based on chance and hence are stochastic. However, the risk of hereditary effects is much lower than the risk of cancer.

Experimental studies on plants and animals have shown that hereditary effects can occur after exposure to large doses of radiation. However, there have been no cases identified in humans where radiation has resulted in hereditary effects.

Hereditary effects from exposure to ionizing radiation have not been identified in humans.

1.3.4. Summary of radiation effects

Acute exposures may lead to early or late effects which may be either stochastic or deterministic. High level chronic exposures could lead to late effects which may be either deterministic or stochastic.

2. Radioactive sources

Radioactive sources are used throughout the world for a wide variety of beneficial purposes in industry, medicine, agriculture, research and education. When such sources are safely managed and securely protected, the risks to workers and the public will be minimized and the benefits will outweigh the associated hazards.

If, however, a radioactive source becomes out of control and unshielded or dispersed as the result of an accident or a malevolent act, for example – persons could be exposed to radiation at dangerous levels. A radioactive source is considered to be dangerous if it could be 'life threatening' or could cause a permanent injury that would reduce the quality of life of the person exposed. The extent of any such injuries will depend on many factors, including: the size of the radioactive source; how close a person is to the source and for how long; whether the source is shielded; and whether or not its radioactive material has been dispersed and caused the contamination of skin or been inhaled or ingested.

In recent years, orphan sources (a source which is not under regulatory control because it has never been so, or because it has been abandoned, lost, misplaced, stolen or otherwise transferred without proper authorization), have caused multiple fatalities or serious injuries when unknowing individuals find them. This problem, along with concern that orphan or vulnerable sources might be acquired for malevolent purposes, has led many countries to consider making concerted efforts. The International Atomic Energy Agency has published many reports providing a methodology as to how to improve control over the sources, related to regulatory infrastructure, emergency response, security, illicit trafficking and border monitoring, and the management of disused sources.

The IAEA categorization (TECDOC-1344) provides a relative ranking of radioactive sources in terms of their potential to cause immediate harmful health effects if the source is not safely managed or securely protected.

3. Malicious uses of radioactive sources

In some countries, regulatory control of radioactive sources – used extensively in medicine and industry – remains weak. Global concerns about the security and safety of radioactive sources escalated following the September 11/2001 terrorist attacks in the United States. Data from the US Nuclear Regulatory Commission in the USA indicate that they have about 150,000 licensees who possess about two million devices containing radioactive sources. Of these licensees, 135,000 are general licensees for the lower categories of sources and about 20,000 are specific licensees for the higher activity sources. The latter are being used in applications such as brachytherapy, teletherapy, industrial radiography, well logging, and laboratory research. In this specifically-licenced group there are about 260,000 devices. NRC data indicate that an average of 375 sources or devices of all kinds are reported lost or stolen each year. Although this is only about 0.02% of the total inventory, it is still approximately one source per day. However, the majority of these are very low activity sources.

3.1. RADIOACTIVE DESPERSAL DEVISES

There are fears that some radioactive sources could be used by terrorists as radiological dispersal devices (RDD's), or so-called "dirty bombs." The radioactive material dispersed, depending on the amount and intensity, could cause radiation sickness for a limited number of people nearby if, for example, they inhaled large amounts of radioactive dust. But the most severe tangible impacts would likely be the economic costs and social disruption associated with the evacuation and subsequent clean-up of contaminated property. A dirty bomb is a conventional explosive that disperses radioactive material. The intent is death and disruption. Death may be the result of the explosive and could result from sufficient radioactive material. Disruption is the result of panic caused by fear of radiation and the cleanup of the radioactive material and any consequent avoidance of the radioactive location after clean up. The amount of radioactive material is irrelevant to disruption. Something that is contaminated will introduce some degree of denial of access, be it a post office, courthouse, subway, or street corner. Dispersal of material is the obvious outcome of an explosion. There are big and little pieces, dusts and smokes. There is a lot of all of it in a relatively small volume trying to move away from the center of mass of the explosion. The direction it goes will depend upon its mass, the relationship to other masses, and the process by which it became a big piece, a little piece, dust, or smoke. An ideal sphere with an explosive in the center, suspended well above the earth with no wind will spread with nearly spherical symmetry. The hot gas will cause a slight upward deflection. A second object placed near the sphere may substantially affect the shape of the explosion, depending upon its mass, shape, and mechanical properties. The simple laws of motion apply, but it takes a super computer to apply them. Maximum dispersal returns us to the ideal sphere with the material to be dispersed outside the explosive. It is also desirable that the material be powdered or highly friable. Maximum dispersal then calls for a bit of engineering. Satisfaction with less than maximum still may require engineering if the material is dense and of a difficult shape.

3.2. SOURCES TO BE CONSIDERED

It is assumed that covert radioactive shipments serve two purposes; they provide material for radioactive dispersal devices or for nuclear weapons. Radioactivity dispersal devices are explosive devices that spread radioactive material for purposes of contaminating a large area and exposing a large number of people to both internal and external radiation doses. These devices need not cause harm to health, but have sufficient potential for harm that costly cleanup is required and access to areas, materials, or buildings is denied. An effective device must have sufficient impact to cause harm or denial. A device of 10^{12} Bq or more would be required, but a 10^{11} Bq source would cause considerable consternation. The materials considered are those that are readily available for medical or commercial uses. They are listed in Table 1 by type of radioactive emission:

TABLE 1. Available radionuclides for medical or commercial uses.

Gamma	Beta	Alpha
Cobalt-60	Phosphorus-32	Americium-241
Iodine-131	Strontium-90	Plutonium-239/238
Cesium-137		Uranium
Iridium-192		Thorium

Material for nuclear weapons is necessarily fissile material, which is generally limited to Uranium-235 and Plutonium-239. There are other fissile materials, but they are both uncommon in sufficient supply and have particular difficulties for forming a nuclear device. Very small nuclear devices are possible using small amounts of fissile material. It is possible that a single individual could assemble a fissile device in less than a week that is transportable by small sedan. Nevertheless, this presumes sufficient knowledge, access to sophisticated electronics, and sufficient fissile material. A device constructed with a small amount of fissile material requires increased sophistication in knowledge and materials as the amount of fissile material decreases. It is assumed that the minimum mass of fissile material is 1,000 g or 2.2 lb. The assumed package weight for shipment is 150 lb or less and the assumed legal package may read 1 μSv/h at the package surface (naturally occurring radiation as in some ceramics). Gamma or x-ray radiation will be the basis for detection. The assumed number of shipments to smuggle sufficient material for either a dispersive or a nuclear device is 10 or less.

It is an ongoing process, attempting to anticipate intelligent use of RDD and assessing terrorist trends. Millions of radiological sources are available, but a large majority of sources (e.g., nuclear medicine diagnostic dosage) are relatively harmless. Usage of an RDD may trigger panic out of proportion of true risk to human health and safety. Possible strategies might be:

- Minor sources – public education and response are crucial

- Large sources – need to tighten control, detect during transport, and prepare appropriate responses

There are basically two broad categories of radioactive material candidates for RDD porpoises.

– Materials that are largely under control at a limited number of sites:
  - Nuclear weapons materials – excess/retired or waste streams
  - Nuclear power related materials – fresh or spent fuel, wastes

– Radiological materials under limited controls and at numerous locations:
  - Industrial, medical, or other applications
  - Many are used worldwide, under variable regulation

It is clear that activity levels and potential radiological dose hazards may vary by many orders of magnitude depending on the materials used to construct RDD, its design, application, weather conditions and many others.

Meteorological data is of great importance in assessment of possible RDD accidents. Assessments should be used as first estimate of the threat and followed by real measurements on the site as soon as possible in the case of a real accident. Field information should be inputted into the scenario in order to obtain more realistic data needed for decision making. This process should be iteratively performed during emergency.

Various data is needed to perform radiological assessment and action plane in the case of RDD explosion. Beside profiles of the inhabitants of the area of interest, their eating and overall living habits, geographical data and weather conditions information for the all four seasons are of paramount relevance. Various dispersion and fallout models are used to perform calculations with various intents. Gaussian Plume models are used to perform first estimates and initial predictions because they require minimal input (just constant wind). Field data is used with more complicated models (Monte Carlo/Gaussian-puff atmospheric dispersion model with vertical variation in meteorological data, 3D Monte Carlo particle dispersion model, Nuclear explosion local fallout models, ...) due to the time needed for models to perform calculations.

Data from the US Nuclear Regulatory Commission in the USA indicate that they have about 150,000 licensees who possess about two million devices containing radioactive sources. Of these licensees, 135,000 are general licensees for the lower categories of sources and about 20,000 are specific licensees for the higher activity sources. The latter are being used in applications such as brachytherapy, teletherapy, industrial radiography, well logging, and laboratory research. In this specifically-licensed group there are about 260,000 devices. NRC data indicate that an average of 375 sources or devices of all kinds are reported lost or stolen each year. Although this is only about 0.02% of the total inventory, it is still approximately one source per day. However, the majority of these are very low activity sources.

4. Contamination of food or water supplies with radioactive material

The deliberate act to contaminate food or water supplies with radioactive materials is unlikely to lead to the significant internal contamination of a large number of people due to the enormous quantities of radioactive material that would be required to reach high levels of contamination in mass-produced or distributed supplies. However, past experience seems to show that the public concern generated by such an act could present a significant challenge to the authorities. Although it is unlikely (though not impossible) that there would be a need for emergency monitoring of a large number of people for internal contamination, there is a need to co-operate with radiological experts and media specialists to quickly assess the potential medical impact of such acts and provide public information to alleviate fears in the potentially affected public. There is also a need to develop a plan, at the national level, to monitor a representative sample of the potentially affected population to confirm the health risk assessment and reassure the public.

In ICRP (International Commission on Radiological) Publication 82, the Commission recommended the establishment of *intervention exemption levels* for commodities including foodstuffs. Commodities that are above such intervention exemption levels should be 'intervened' and those that are below could be 'exempted'. The Commission noted, moreover, that mainly due to the globalization of markets, intervention exemption levels of radionuclides in commodities may not be amenable to ad hoc decisions, i.e. they cannot be established on a case-by-case basis but rather need to be standardized internationally. It is not likely that several types of commodities would be simultaneous sources of high exposure to any given individual. On the basis of this presumption, the Commission recommends a generic intervention exemption level of around 1 mSv for the individual annual dose expected from a dominant type of commodity, such as some building materials that may in some circumstances be a significant cause of exposure. It should be noted that commodities produced or subject to commerce within the area of influence of the malevolent event would present an exceptionally difficult situation.

If the corresponding activity levels are higher than those in produce from neighbouring areas, issues of market acceptance could arise – particularly if there are transboundary movements of the commodities.

Following the Chernobyl accident and taking account recommendations of the Commission at the time, the Codex Alimentarius Commission adopted generic intervention exemption levels for radionuclides in foodstuffs following a nuclear accident. In response to the new recommendations of the Commission, these levels have been revised by the Committee on Food Additives and Contaminants of the Codex Alimentarius Commission [CODEX ALIMENTARIOUS 2004]. The levels are:

- 10 Bq/kg (becquerels per kilogramme) for 238Pu, 239Pu, 240Pu and 241Am
- 100 Bq/kg for 90Sr, 106Ru, 129I, 131I, and 235U
- 1,000 Bq/kg for35S, 60Co, 89Sr, 99Tc, 103Ru, 134Cs, 137Cs, 144Ce, and 192Ir
- 10,000 Bq/kg for 3H, and 14C

The Commission considers that the revised levels should be applied to manage the distribution of contaminated foodstuffs following a radiological attack, specifically if international trade is involved. If the annual doses in the area affected by the event are acceptable because the intervention strategy has been optimized, the situation outside the affected area should also be acceptable because the individual annual doses elsewhere from the use of commodities produced in the affected area would normally not be higher than those in the event have not been lifted, production of the restricted commodities should not be restarted; conversely, if the restrictions have been lifted, production can be restarted.

If an increase in production is proposed, it could proceed subject to appropriate justification. In circumstances where restrictions have been lifted as part of a decision to return to 'normal' living, the resumption and potential increase of production in the affected area should have been considered as part of that decision and should not require further consideration.

Bibliography

International Atomic Energy Agency, Safety of Radiation Sources and Security of Radioactive Materials, (Proc. Int. Conf., Dijon, 1998), IAEA, Vienna (1999).

International Atomic Energy Agency, National Regulatory Authorities with Competencies in the Safety of Radiation Sources and the Security of Radioactive Materials, (Proc. Conf. Buenos Aires, 2000), C&S Papers Series No. 9/P, IAEA, Vienna (2001).

International Atomic Energy Agency, International Conference on the Security of Radioactive Sources, Findings of the President of the Conference, IAEA, Vienna (2003).

International Atomic Energy Agency, Action Plan for the Safety of Radiation Sources and Security of Radioactive Materials, GOV/1999/46-GC(43)/10, IAEA, Vienna (1999).

International Atomic Energy Agency, Categorization of Radiation Sources, IAEA-TECDOC-1191, Vienna (2000).

International Atomic Energy Agency, Strengthening Control over Radioactive Sources in Authorized Use and Regaining Control over Orphan Sources, National strategies IAEA-TECDOC-1388, Vienna (2004).

Meserve, R.A., "Effective Regulatory Control of Radioactive Sources", National Regulatory Authorities with Competence in the Safety of Radiation Sources and the Security of Radioactive Materials (Proc. Int. Conf., Buenos Aires, 2000), IAEA-CN-84/2, IAEA, Vienna (2001).

ICRP Publication 82, Protection of the Public in Situations of Prolonged Radiation Exposure, 82, Annals of the ICRP Volume 29/1–2, Elsevier, 2000.

Codex Alimentarius Commission, Joint FAO/WHO Food Standards Programme, Fourteenth Edition World Health Organization Food and Agriculture Organization of the United Nations, Rome, 2004.

Charles D. Ferguson et al, Commercial Radioactive Sources: Surveying the Security Risks, ISBN 1-885350-06-6, Monterey Institute of International Studies, 2003.

AN EXPLORATION OF POTENTIAL CHEMICAL
AND BIOLOGICAL THREAT AGENTS

CHRISTINE HANSEN

1897 Beach Road RR#2, Oxford Mills,
Ontario, KOG 1SO, Canada
e-mail: Chris.Hansen@inspection.qc.ca

Abstract: Food and water provide an attractive and convenient target for intentional contamination. Food and water often follow complex production, and supply systems and the technical, physical and operational barriers are unbounded as compared to other types of potential targets. Globalization also plays a large role in making this target more appealing with the rapid and widespread distribution of food, and food ingredients from many locations providing potential for a greater dispersal of an agent. In short, there is potential for significant vulnerabilities all along the food and water continuum. Another factor that contributes to the attractiveness of food and water as a target for intentional contamination is the fact that often in the initial stages of a response to a foodborne illness outbreak related to food contamination; it could be very difficult to distinguish the intentional from the accidental. This potential lag in reaction time to the intentional contamination may facilitate an ongoing and wider spread of the agent in addition to providing the perpetrator with the opportunity for to elude capture.

The attributes that make food and water very attractive as a target can also make it a very challenging target. Much of the food that we eat is not a single component but often a blend of components. Each matrix is slightly different and each potential agent may react differently when exposed to the matrix. Due to the complex nature of food and food systems, the advancing technologies in genome sequencing etc. it is difficult to create a finite list of potential chemical and biological threat agents especially when discussing potential terrorism threats to food and water. In this presentation we will review some of the history of intentional contamination of food, and explore some of the characteristics of biological and chemical threat agents that would make them attractive for use as possible weapons against a given population.

Keywords: Biological agents, categorization, chemical agents, food terrorism

1. A historical review

There is a long history of biological and chemical agents being used to contaminate food and water systems as acts of war or terrorism. Throughout human history there are reports contamination particularly of water sources. For example, in the sixth century BC, ergot (*Claviceps purpurea*), a fungus found in rye was used by the Assyrians to poison the wells of their enemies.[1] The ergot fungus creates alkaloids that have an effect on body functions such as circulation, sometimes leading to gangrene and neurotransmission causing a range of symptoms from irrational behaviour to death. The common name for ergot related disease is "St Anthony's Fire" (med terms.com "St. Anthony's fire- ergotism). Other agents used in early times include hellebore roots and human or animal carcasses in water.

In modern times, the use of chemical and biological agents in foods has continued and increased in sophistication:

- In 1978, Israeli Jaffa oranges destined for Europe (Holland, Germany, Britain) were poisoned with metallic mercury. The attackers wanted to economically affect the Israeli orange export market by injecting oranges with mercury. While no deaths were attributed to this attack at least a dozen people were ill, and the attack could be considered successful as it resulted in a reduction in the exports of oranges by approximately 40% in that year.[2]

- In 1984 members of the Rajneesh Cult in Wasco, Oregon USA tested the use of *Salmonella* in restaurant salad bars. The intent was to influence local elections. As a test of the system several weeks before the elections, Rajneeshees members armed with multiple vials of *Salmonella* employed it by discreetly pouring the solution into salad bar bowls, salad dressing bottles and coffee creamer containers in ten restaurants.[3] Following the attacks 751 people fell ill with salmonella gastroenteritis, 45 of which had to be hospitalized. The overall objective of the plot which was to influence the outcome of the elections was thwarted by electoral officials in Wasco. The actual planned attack for Election Day never occurred.

- In 1995, Tajik opposition members laced champagne with cyanide at a New Year's Eve celebration. The attack resulted in the deaths of seven people and sickened many others.[4]

- In 1998, arsenic was used in curry served at a Carnival in Wakayama Japan. Four people died and 75 were sent to hospital. According to reports, the original agent was presumed to by cyanide and it took almost a week to confirm arsenic. The alleged attacker was a woman, an insurance sales person, and the presumed motive was anger at her neighbours for shunning her.[5]

- In 2003 there was the "beer and burger plot", an al Qaeda terrorist plotting a bomb attack on Britain told accomplices to get a job at a stadium and sell contaminated beer at soccer games "You just put poison in a syringe, inject it in a beer can and put a sticker on it, which would stop it leaking, and hand them out," the same terrorist also suggested getting "a mobile vending cart selling burgers, just poison those. You could set up a shop on a street corner and sell poisoned burgers and then all you have to do is leave the area."[6]

- In 2008 in Iraq, thallium (thallium sulphate probably) an odourless, tasteless chemical sometimes used in insecticides and rat poison was put into a cake that was delivered to a sports club. Some of the sport club officials ate the cake and some and took some home to their families. The poisoned cake was sent by a sports coach who had left the club on bad terms. As a result of this poisoning, two children died and five people were seriously ill.[7]

More recently there have been some very disconcerting examples of intentional contamination of a food or food component or ingredient, which clearly demonstrates the current vulnerability of the global food system. While the following examples are more appropriately classified as intentional fraud, and not terrorism, it may be of value for risk profilers and vulnerability assessors as it clearly demonstrates several areas of vulnerability, and perhaps draws a bit of a roadmap for future attackers.

In spring of 2007 there was a massive intentional contamination wheat gluten used as an ingredient in pet food with a fertilizer called melamine (Melamine is a molecule that has a number of industrial uses, including use in manufacturing cooking utensils. It has no approved use in human or animal food in the United States or Canada, nor is it permitted to be used as fertilizer, as it is in some parts of the world). While it is difficult to determine exact numbers of pet deaths that occurred as a result of this intentional addition as there is no centralized government records database of animal sickness or death in North America as there are with humans. Conservative estimates place the potential number of deaths above 10,000 animals. The contamination resulted in a recall of more than 70 different brands of pet foods. It is generally believed that the intentional addition of the melamine was more an issue of fraud than an intent to kill animals. However, this incident clearly demonstrates the vulnerability of food ingredients sourced from outside of the country. It also demonstrates the vulnerability of some food systems as the key recalling firm was a large co-packer who while they created different formulas for the various brand names, used the same base ingredients in all the formulations.

Ironically, in fall of 2008, a similar concern with the same chemical, melamine, had surfaced again. On September 12, 2008 there were reports from China of melamine contaminated infant formula. According to media reports more than 54,000 infants and young children have sought treatment for urinary problems, possible renal tube blockages and possible kidney stones related to the melamine contamination of infant formula and related dairy products. Three deaths among infants have been confirmed, more than 14,000 infants have been hospitalized and a little less than 11,000 remain sick.[8] Contamination has also been found in liquid milk, shelf stable milk products, frozen yogurt dessert, biscuits, candies and in coffee drink. All these products were most probably manufactured using ingredients made from melamine contaminated milk. In this event, contamination appears to have happened as fraudulent contamination in primary production. Chinese government officials have pinpointed milk collecting stations as the sites where the melamine was added. The then contaminated milk was used in the manufacture of powdered infant formula processed before 6 August 2008 and the tainted milk powder has also been used in the manufacture of a number of other products that were shipped around the world causing many countries to initiate recalls of potentially contaminated products.

If one reviews several of the more recent large scale food outbreak situations in Canada and the USA, there may also be some valuable lessons for example: the potential movement of foodstuffs especially concerning the speed of distribution and the number of locations; the wide usage of certain common ingredients and additives in a variety of different foods; the challenges in performing trace backs and trace outs of foods.

2. The challenge

In a system as complex as the food continuum, it is nearly impossible to tackle all hazards related to potential intentional contamination. The challenge for those who are engaged in food defence is to narrow down the list of potential threat agents based on their characteristics in an attempt to focus both recourses and efforts on the most likely agents. For example there are approximately 30,000 chemical agents that have some level of toxicity to humans. It would be impossible and impractical to develop, validate and perform complex chemical testing in a variety of food matrixes for all of these agents.

To add to the difficulty of this task, in food defence, threat agents cannot generally be considered in isolation as their intended use is in a complex matrix of carbohydrates, fats, oils, food additives, and proteins. There is a potential for chemical interactions, and in the case of biological agents many organisms require specific environments to survive and they will not thrive in particular environments (i.e. low water activity, acidic, high salt etc.)

and these situations may assist in defeating the threat. In order to develop a robust food defence system, it is important to link the threat agent to the target food matrix as closely as possible as the complexity of the food, the potential amount of agent required to contaminate the food, or the food processing itself may neutralize the attack.

As mentioned previously, it is very difficult to create a finite list of threat agents. The potential choice of a threat agent has many variables. The choice of agent may be influence by the overall objective of the attacker in some cases the attacker may be satisfied with causing mass illness, other times the objective is to kill as many as possible, in another situation the disruption and fear may be enough. One should also consider the fact that the use of some threat agents may be more country specific depending on the potential availability of the agent or precursors, and/or the availability of the technology required for producing sufficient amounts of a substance, chemical, or biological to contaminate food at a dose large enough to cause significant illness or lethality, as well as the accessibility to the target.

3. Chemical threat agents/contaminants

Chemical agents could be used to cause death, incapacitation, or disruption to a targeted population. Chemicals agents include those that were developed for use in the battlefield as well as toxic industrial chemicals, pesticides and chemical food additives.

1. As mentioned previously, there are thousands of chemicals that are considered toxic to humans; however, not all of these chemicals would be appropriate for use to contaminate food. For example, the US CDC has a list of approximately 88 chemical agents[9] that can be a threat to humans. Approximately 19 of the substances on that list are suitable for possible ingestion, and the number of potential agents gets further reduced when considering the potential interaction of the food matrix and the chemical agent.

There are certain characteristics that will make a chemical agent more attractive for use in an attack on a food:

- Acutely toxic through ingestion – for terrorism purposes, the main objective is to kill or injure or, incapacitate as many people as possible in a relatively short period of time. Agents that are acutely toxic (i.e. have an LD_{50} of less than 50 mg/kg oral) require less of the agent to successfully impact many individuals. Also, the flexibility to use smaller amounts of the agent possibly carried in a vial or a small bag will reduce the potential of being caught in the act of contaminating the food.

- Odourless – if a particular agent is odourless or has only a slight odour, the chances of being detected in the food are minimized. Those agents that may have a slight odour can possibly be used in more pungent food preparations, for example in a curry sauce.

- Tasteless – if a chemical agent cannot be tasted, it can be used in a wider variety of foods. As with odour, some agents with slight tastes can be used in certain food preparations and remain undetected.

- Solubility – products that have greater levels of solubility have the capability to be more evenly mixed into a food matrix and thus disguised this will reduce the chance of detection in the food matrix during quality control checks etc.

- Stability in the food matrix – the agent must be stable in the food matrix so that maximum toxicity is maintained. Reactivity with various chemicals in the food may result in a reduction in the toxicity of the agent.

- Easily available or produced commercially – in general, agents that are readily available in nature (heavy metals) or produced commercially are more likely to be used. The accumulation of large amounts of such agents may escape detection by authorities. The attackers may also target High Production Volume (HPV) chemicals such as pesticides, industrial chemicals, food additives, which are chemicals produced or imported to Organization of Economic Cooperation and Development (OECD) countries in excess of 1,000 t per year.

- Limited technical expertise or no purchasing controls – agents that are easily synthesized with limited technical knowledge and expertise and agents or precursors that are readily available without special permits are more attractive to terrorists. This characteristic facilitates the purchase or production of potentially large amounts of an agent without the risk of detection by authorities.

- Potential for mass fear – there are some agents that can create fear in the public and therefore have certain attractiveness to a terrorist. Public perception is something that is hard to predict and is likely to change over time depending upon the specific experiences of the population. Agents with higher mortality will likely create more fear. The fear in the population may also be augmented if the agent used is not usually found in food.

- Difficult to detect in food – when the agent is difficult to detect or isolate in food, there will likely be a delay in identification of the contaminated food by traditional food safety procedures which has a cascading effect thus delaying any mitigating activities such as recall that would halt the distribution of affected food. This characteristic will potentially allow more people to consume contaminated food for a longer period of time.

- Difficult to diagnose in humans – difficulty to diagnose may delay effective treatment and thus may have a greater impact on the targeted population. In the case of the contamination of curry mentioned earlier in this paper, the medical officials originally diagnosed the symptoms as cyanide poisoning and the wrong antidote was administered to victims initially.

- Difficult to treat in humans – if the disease cannot be treated then the impact on the human targets may be greater.

- Historical examples of the agent being used to contaminate food – if there are historical examples of a chemical agent being used in food there is a greater likelihood of it being used again for example arsenic, which has a long history of being used to poison people in various situations.

4. Biological agents/contaminants

Many pathogens and toxins cause diseases or have toxic effects in humans. For the purpose of this discussion biological agents will include pathogenic bacteria, viruses, protozoa, prions and toxins derived from biological sources. Many governmental organizations have put together lists of biological agents that are of significant concern to human health and these lists do not change radically when we consider biological agents that could be used in an attack.

Biological agents may be prepared as a fine powder, or liquid. With the exception of acute levels of certain toxins, biological agents produce no immediate symptoms in exposed individuals. Biological agents are too small to see and the release may be invisible. These characteristics make it hard to determine that an attack has occurred. The broad array of possible agents makes them hard to detect. Pathogens also cause varying symptoms and medical expertise is required in order to interpret the symptoms and to order the appropriate confirmatory tests.

There are certain characteristics[10,11] that will make certain biological agents more attractive than others:

- Transmissible through ingestion – agents that can be delivered using food and or water.

- Stability – the more attractive agents will be those which are more resistant to environmental conditions and potential changes in environmental conditions, and remain viable to cause human disease or toxins that resist denaturisation to remain toxic at the point of delivery.

- Availability – agents that are more readily available are more likely to be used by bioterrorists. There are many common agents that cause serious diseases in humans and these agents have natural reservoirs and therefore do not require purchasing from culture collections to procure.

- Limited technical expertise and equipment required to propagate or amplify – agents that require very little technical expertise or knowledge to propagate and or amplify, and unsophisticated equipment for propagation or amplification are more likely to used. An example of this is the use of salmonella in the Rajneesh incident, mentioned previously, where cultists were able to obtain salmonella and other pathogens and produce significant amounts of the bacteria in a small lab. In this case the amount of expertise required was that of a medical technician.

- Highly pathogenic/low infectious dose in food – an agent with a lower infectious dose through ingestion will require fewer organisms to cause disease, and may be potentially more difficult to detect in foods by current laboratory methodologies. If the laboratories need to enrich and amplify the sample, it is likely that the contaminated product will remain in the marketplace longer, thus affecting more people.

- High mortality – in general, those agents that cause fatality will increase the potential impact of the attack, potentially resulting in higher levels of panic in the population.

- Difficult to detect in food – if the laboratory has trouble detecting or isolating the agent then potential mitigating actions such as recall may be delayed, the product remains available to people longer and thus the potential for more people to be affected increases.

- Potential for mass fear – there are some agents that will create fear in the public and therefore have a certain attractiveness to a terrorist. Public perception is something that is hard to predict and it will change over time depending upon the specific experiences of the population. Agents with higher mortality will create more fear. The fear in the population may also be augmented if the agent used is not usually found in food. Fear in a population will also undermine the confidence of the people in existing food safety and food emergency response systems.

- Historic examples – if an agent has been used before in an attack or an intentional contamination of food, it is more likely to be used again as its efficacy has been proven. The previous uses can serve as a blue print for future attacks.

- Difficult to diagnose in humans – the longer it takes to identify the agent the longer it takes to provide the appropriate treatment to the victims which may result in greater numbers of causalities.

- Potential for human–human transmission – this will potentially result in additional casualties for those people who come into contact with persons infected by the food, ultimately causing a greater burden on health care officials. This may also slow down the identification of the contaminated food since sick individuals did not ingest the suspect food. Problems in identifying the contaminated food will slow mitigating actions such as recall.

- Genetic modification – genetic modification can increase the virulence of an agent or impart some level of drug resistance making it more difficult to treat and possibly resulting in greater casualties.

- Difficult to treat in humans – if a disease does not have an effective treatment, the number of casualties will increase.

5. Conclusion

It is highly unlikely that there will ever be a complete and finite list of chemical or biological agents that could be used in a bioterrorism attack on food or food systems. Lists that are created should be refreshed and validated on a regular basis as the landscape changes constantly. We know that potential attackers are growing in their creativity and technical capacity, and new chemical and biological agents are discovered and/or developed regularly.

The attributes that make food and water an attractive target also make food a challenging target. There are literally thousands of different combinations of food and agents that could potentially be used in an attack on food. It is useful however, to understand the characteristics that make certain potential agents more attractive for use than others. This understanding linked with a good understanding of vulnerable foods is useful for those who battle against food terrorism in order to better focus resources and anticipate potential attacks.

Bibliography

1. James A. Romano, Brian J. Lukey, Harry Salem: *Chemical Warfare Agents* 2007.
2. Byconederbyshire.ed.uk
3. Thompson, Christopher: The Bio-Terrorism Threat by Non-State Actors: Hype or Horror". A thesis published by the Naval Postgraduate School, California, 2006 pp.20-30 http://www.ccc.nps.navy.mil/research/theses/thompson06.pdf
4. Cote, Francois, Genevieve Smith, *Chemoterrorism* Government of Canada Publication 15, January, 2002
5. BBC News Word Edition, "Japan's Curry Killer Sentenced to Death"Wed 11 December, 2002
6. CBS News "Qaeda Tested Poison Beer and Burger Plot" March 25, 2006
7. CDC MMWRH September 19, 2008 Thallium Poisoning from Eating Contaminated Cake – Iraq 2008
8. WHO – October 2008 – Melamine-contamination event, China, September–October 2008

9. http://emergency.cdc.gov/agent/agentlistchem-category.asp
10. MacIntyre, C Raina: Development of a Risk-Priority Score for Category A Bioterrorism Agends as an Aid for Public Health Policy. Military Medicine, 2007
11. Retez, Lisa D Ali S. Khan, Scott R. Lillibridge, Stephen M. Ostroff, and James M. Hughes Public Health Assessment of Potential Biological Terrorism Agents Centers for Disease Control and Prevention, Atlanta, Georgia, USA Vol. 8, No. 2

FOOD QUALITY SYSTEMS IN TURKEY: PERSPECTIVES IN TERMS OF FOOD DEFENCE

HAMİ ALPAS

Food Engineering Department, Middle East Technical University, 06531, Ankara, Turkey
e-mail: imah@metu.edu.tr

SERHAT AŞCI AND AHMET ALI KOÇ

Department of Economics, Akdeniz University, Antalya, Turkey
e-mail: serhatasci@akdeniz.edu.tr; alikoc@akdeniz.edu.tr

Abstract: The chapter summarizes the current food quality systems in Turkey from the perspective of food defense. Recently Turkey has formally adopted a number of typical elements of food safety regulations and control systems in the accession period to EU and there are developments dealing with food safety and a few available empirical analyses of food safety applications in Turkey, but still there is a lack of comprehensive study summarizing the efforts. The developments that signal some of the more formal approaches to deal with food defense are shared and a few examples of food safety applications in Turkey are mentioned together with discussing issues about food quality. The policies supported for food safety and security, current situation, related legislations are used to provide information as current indicators. On the other hand food defense is a vital and relevantly "hot" topic as all societies are crucially dependent upon the food supply; therefore, its disruption is an obvious prime target for terrorism.

Keywords: Food defense, quality systems, terrorism, Turkey

1. Introduction

Food quality is the quality characteristics of food including external factors (texture, flavour, origin and appearance; size, shape, colour) and internal factors (chemical, physical, microbial) (Grunert 2002). Food quality also

V. Koukouliou et al. (eds.), Threats to Food and Water Chain Infrastructure,
DOI 10.1007/978-90-481-3546-2_3, © Springer Science + Business Media B.V. 2010

deals with product traceability of the raw materials, ingredients and packaging suppliers as consumers may be susceptible to any form of contamination and also they require trust on manufacturing and processing standards. In addition, food quality also deals with labelling issues to ensure there are correct product, ingredient and nutritional information.

Agriculture still plays an important role in Turkish economy, even though its share in the economy has decreased significantly during the last few decades. The agricultural sector made up about 22% of Gross Domestic Product (GDP) at the beginning of the 1980s, but that has declined to less than 10% in recent years. It is still an important buffer against urban unemployment. However, nearly 30% of the economically active population lives in rural areas (SPO 2007), while agricultural employment accounted for 24.6% of all employment in December 2008, according to the participation of the workforce (TUIK 2009). Agriculture, fishery and food products-beverage made up around 7.8% of export value of US$124.3 billion between January and November 2008 (SPO 2007).

The size of Turkish food sector is estimated to be €45 billion globally. Food sector represents a 20% share in total production of the manufacturing sector and contributes approximately 5% to Gross National Product (GNP) (Guittard 2006). The food sector employs more than 250,000 registered workers and technical staff in nearly 30,000 enterprises. Most of them are small to medium-sized enterprises. The State Planning Organisation (SPO) estimates that around 10% of these enterprises are relatively modern and large. USDA (2004, GAIN TU#2008) reports that only 17% of these firms use proper quality control tools and only a small proportion of these firms meet the EU quality norms (Oskam et al. 2004).

Policies dealing with food safety and quality started to develop in the mid-1990s due to the custom union with the European Union (EU) in 1995 and strengthened during 2000s, because of export to developed market economies which require greater concern of food safety and quality. The penetration of supermarkets into domestic retail markets is another driving force behind food quality and safety (Oskam et al. 2004): Simulations of the long-term impact of EU accession suggests that the increase in market access into the EU could generate a significant increase in demand in both quantity and quality that would support significant growth of the agricultural and food sectors in Turkey (World Bank 2006). Turkey has formally adopted a number of typical elements of food safety regulations and control systems in the accession period to EU. There are developments which signal some of the more formal approaches to deal with food safety and a few available empirical analyses of food safety applications in Turkey, but it is difficult to discuss food quality issues as there is no accessible comprehensive study and to define quality is more difficult.

This contribution provides information about the current situation on food quality systems in Turkey from food security perspective. Using the FAO definition of food security:

"Food security exists when all people at all times have physical and economic access to sufficient, safe and nutritious food to meet their dietary needs and food preferences for an active and healthy life style" (FAO 1996). The addition of "safe and nutritious" emphasize food safety and nutritional composition while the addition of "food preferences" changes the concept of food security from mere access to enough food, to access to the food preferred (Pinstrup-Andersen 2009). This implies that people with equal access to food, but different food preferences, could show different levels of food security.

1.1. PUBLIC INSTITUTIONS

1.1.1. Ministry of Agriculture and Rural Affairs

The Ministry of Agriculture and Rural Affairs (MARA) include two essential sections which are main service units and advisory and control units. The main services units of MARA are the General Directorate of Agricultural Production and Development, General Directorate of Plant Protection and Control, General Directorate of Structuring and Support, General Directorate for Agricultural Research and Department of Foreign Affairs and EU-Coordination. MARA conducts research, prepares plans and programs on the improvement of agricultural production, conservation of natural resources such as land, water, plants and animals. In addition, support of animal breeding, control of labelling, food and feed production and usage of plant and animal drugs, supervision of services related with food and feed, control of animal diseases, provision of agricultural services and infrastructure, and rehabilitation of social services related to agriculture are in the mandate of MARA (www.tarim.gov.tr).

In Turkey, MARA is generally carrying out the food safety responsibilities. According to the Food Law (no: 5179, published in official gazette as of 5 June 2004, no: 25483), MARA mainly concerns the technical and hygienic aspects in food production places and focuses on the issuing of production licences for producers of foodstuffs and the control of sale and consumption points of foodstuffs. Besides, MARA also concerns the issuing of permission and control of import and export foods. In addition, MARA is the contact point of Codex Committee in Turkey and related product communiqué studies (Turkish Food Law No 5179, 2004).

The food inspection system of MARA is composed of 40 laboratories within 81 provincial directorates and approximately 5,000 food inspectors (including main, assistant and authorised inspectors) are carrying out food inspection (www.tarim.gov.tr).

MARA is also responsible for the general management of the organic fruit and vegetable cultivation system, good agricultural practices (GAP) and biotechnology-biosecurity in Turkey. The Secretariat of Organic Agriculture and the Organic Agriculture Committee was established in 1993 under the Department of Research, Planning and Coordination, which is classified as one of the advisory and control units. Through a decision of the Minister, the responsibility was taken from the Department of Research, Planning and Coordination and given to General Directorate of Agricultural Production and Development in 2003 and Good Agricultural Practices[1] (ITU) section was also established at the same unit. Currently staffs dealing with issues on organic agriculture and good agricultural practices are employed at the Department of Alternative Agricultural Production Techniques. Biotechnology sub-commission, established under the control of General Directorate for Agricultural Research at MARA, was composed of academicians, public officials and NGOs to adopt EU regulations to Turkish regulations like GMO and novel foods.

1.1.2. Turkish Standards Institution

Turkish Standards Institution (TSI) has been established by the law number 132 on 18 November 1960 for the purpose of preparing standards for every kind of item and products together with their procedure and service. Turkey has been a member of ISO (International Organisation for Standardization) since 1955.

The Institute is a public founding conducted according to the special rules of law and has a juristic personality. The standards that have been accepted by TSI get the name of Turkish Standards. These standards are voluntary and can be made compulsory by the approval of the ministry that the standard is relevant to. It is essential that a standard be a Turkish one so that it could be made compulsory. The standards made compulsory are published in official gazette. If a firm desires to be accredited by ISO: 9000 or ISO: 22000, Turkish Standards for products published by TSI have to be applied and it should be audited by private auditors from certification firms. There are almost 1,750 standards prepared by TSI related with food speciality, quality and sanitary aspects. All the standards will be harmonised with EU standards and almost 600 standards have been repealed by new standards since 1990s. However, there is not any mandatory domestic market inspection executed by TSI, although TSI contains certification auditor for voluntary product,

---

[1] ITU refers to Good Agricultural Practices (GAP) in Turkish and the ITU standards are regulated by Turkish legislation and it could be called as TurkishGAP.

production and service places, laboratories and vehicles (www.tse.gov.tr). Furthermore, all the goods imported and exported must be compatible with TSI standards and this is inspected by Undersecretariat for Foreign Trade (UFT) which is detailed in Section 2.4.

1.1.3. Turkish Patent Institute

Turkish Patent Institute (TPI) was established in 19 June 1994 based on the power granted by the Law No: 4004 dated 16/6/1994, as an independent legal entity with a special budget acting under the authority of Ministry of Industry and Trade, with the objective of supporting the technological development in Turkey and establishing and protecting industrial property rights, as well as providing the public with the worldwide information on industrial property rights, thereby supporting the cultivation of a competitive environment and the development of research and development activities.[2]

The legislations have been created as a reform feature to undertake liability on patent, trademarks, industrial design and geographical indications[3] (GIs), derived from "WTO agreement" and its annex IC "Agreement on Trade-Related Aspects of Intellectual Property Rights" and Custom Unions of European Union. Turkey has been a contractor in 11 international agreements since 1994. Therefore, special courts were established to create powerful industrial property rights throughout the country and important achievements has been performed with educations and presentation functions to enlighten public oriented users of the system.

Moreover, legislation amendments to minimize the differences have been realized in this period as being part of international agreements. In last decades, TPI achieved further improvements on domestic and international relationship with publicity and institutional development areas. In addition, working capacity was also increased.

1.1.4. The Undersecretariat for Foreign Trade

The Undersecretariat of Foreign Trade (UFT) prepared the "Ministerial Decree on the Regime of Technical Regulations and Standardization for Foreign Trade" and its supplementary legislation with the aim of providing transparency in the implementations, assembling all the dispersed regulations regarding standardization policies in Turkey and establishing a legal base for the harmonization of Turkish legislation with the Community's (www.dtm.gov.tr).

---

[2] Detailed information about TPI is available in the Web Page in both English and Turkish (http://www.turkpatent.org.tr)
[3] GIs covers both PDO and PGI according to the Turkish legislation.

The "Decree on the Regime of Technical Regulations and Standardization for Foreign Trade" is in conformity with the requirements laid down in the Agreement on Technical Barriers to Trade of the World Trade Organisation. It prohibits discrimination among trading partners and it aims to ensure that import products comply with the requirements of protection of human health and safety, animal or plant life or health, or the environment.

The Regulation is related with the controls of the agricultural products to be exported within the scope of the standards mandated in exports. The Regulation also determines the framework of the import controls, which are regulated by communiqués in more detail. The aim of this application, dated back to 1930s in Turkey, is to protect the prestige of traditional Turkish agricultural products and create stable markets in foreign countries.

The standards that are mandatory in exports are TSI standards. These standards are parallel to the UN/ECE standards and the inspections are performed according to the OECD scheme. Following the inspection carried out by the inspectors, a "Control Certificate" is given to the exporter if the product is found to be in conformity with the relevant standard. The exporter cannot export the product without a Control Certificate.

The products shall be exempted from inspection if the exporter owns the Certificate of Competence on Commercial Quality Inspection. Certificate of Competence on Commercial Quality Inspection is a certificate issued by the UFT for the producers who are found to be competent to carry out the inspections by themselves. These firms are subject to periodic and random controls by the Inspectorates.

1.2. AUTONOMOUS AND PRIVATE BODIES (SEMI-PUBLIC ORGANISATION)

Turkish Accreditation Agency (TÜRKAK) established in 1998 has 33 employees. TÜRKAK acts as the major building stone of quality in conformity assessment. TÜRKAK has tried to achieve informing and enlightening the private sector, particularly in the industrial regions which are far from the central decision making process. At present, accreditation decisions are taken by an authorized committee composed of the General Secretary and two deputies who are more competent in evaluating accreditation operations than the Board of Management.

Since April 2006 TÜRKAK is a signatory to the multilateral agreements of the European Accreditation Cooperation (EA MLA) for testing, calibration, inspection, QMS certification. TÜRKAK has also been evaluated for product certification, EMS and certification of persons. Since May 2006,

TABLE 3.1. List of Accredited Bodies by TÜRKAK (www.turkak.org.tr. Access date 20 August 2008).

List of accredited firms and institutions	Food related accreditation	Total
Testing laboratories	27	152
Calibration laboratories	–	44
Quality management system	26	37
Inspection bodies	2	30
Product certification bodies	3	6
Personnel certification bodies	2	8

TÜRKAK is also a signatory to the International Laboratory Accreditation Cooperation – Mutual Recognition Arrangement (ILAC MRA) for testing and calibration. In June 2006, TÜRKAK became a member of International Accreditation Forum, (IAF). The list of accredited bodies by TURKAK is given in Table 3.1.

The Hazard Analysis and Critical Control Point (HACCP) concept was first introduced in Turkey with articles 16 and 17 of the TGKY (Turkish Food Codex Regulation) in 1997. These are included in the specific chapter on food hygiene which quite clearly describes the principles and application steps of HACCP to be incorporated in food plants to achieve the hygienic conditions defined. These are foreseen to cover all food sectors, but a period of adaptation has been given to the food industry before mandating it and subjecting it to official inspections. Many food manufacturing and/or retailing companies in different food sectors, particularly food manufacturers exporting to the EU countries since it is compulsory with the Custom Unions, has started to employ HACCP since 1997.

The HACCP system has become compulsory both for domestic market and all export destinations with the regulation enacted as of 30 March 2005 and entitled as "Regulation on Market Surveillance and Control of Food and Food Contact Materials and Responsibility of Food Business Operators" which was repealed by the "Regulation on Inspection and Control of Food Safety and Quality" published on 9 December 2007/26725 (Regulation 2007). All of the food manufacturers must put into place the HACCP quality assurance since 31 March 2008 (Regulation on Inspection and Control of Food Safety and Quality, 2007).

ISO 22000:2005 is a new standard to supply safe food all over the world. The subject of this standard is to make all the requirements common among farmers in farm, transportation, storing operators, retailers, restaurants and suppliers (including machine and food packaging material producers). It is expected that ISO 22000, published in September 2005, will replace all the HACCP standards published separately by various countries and will be accepted as common standard as ISO 9000 (TS EN ISO 22000, 2005). However, the requirements of ISO 22000 to replace standards like BRC and IFS is not actually common. ISO 22000 was published by TSI as TS EN ISO 22000 in October 2006 and replaced TS 13001 standard prepared by TSI in accordance with Danish Standard DS 3027 in 2004 (TS 13001, 2003). The certification process of TS EN ISO 22000:2005 and other quality management systems has been supplied by accredited (from TÜRKAK) private certification firms. In addition, the certification process of BRC, IFS and foreign nations oriented standards (such as GlobalGAP and TNC) is also carried by these certification firms; however they are accredited from foreign accreditation bodies. The food manufacturing firms having quality systems in Turkey is given in Table 3.2.

The producer, manufacturer, supplier, importer and exporter of organic agriculture products (vegetable and animal products, water, seed, fertilizer, seedling, plant and other inputs, foods, vitamins, the additive materials and industrial products from agriculture orient) have to make an agreement with certification firms authorized by MARA. They could not sustain their activities without this agreement.

TABLE 3.2. Food manufacturing firms using quality assurance systems in Turkey (web sites of the firms).

Sub-sector	Certificate types	Capacity	Firms
Meat and meat products	IFS	20,000 t/year	Aytaç
	ISO 22000	20,000 t/year	Aytaç, Baş.yazıcı
	HACCP- TS 13001	3,000–99,500 t/year	Altınkaya, Etsan/Apikoğlu, Şalvarliet, Yılmazlar Et, İkbal, Sultan Et, Namet, Pınar Et, Van Et, Altın Et, Maret
	ISO 14001	99.500 ton/year	Pınar Et
	ISO 9001:2000	3,000–99,500 t/year	Aytaç, Pınar Et, Altınkaya, Etsan/Apikoğlu, İkbal, Sultan Et, Namet, Baş yazıcı, Maret
Boiler	HACCP- TS 13001	50,000–240,000 units/day)	CP Gıda A.Ş., Banvit, Beypiliç, Er Piliç, Şen Piliç, Keskinoğlu, Şeker Piliç, Köy-Tur, Emre Piliç.

(continued)

TABLE 3.2. (continued)

Sub-sector	Certificate types	Capacity	Firms
	ISO 14001	240,000units–250 t/day	CP Gıda A.Ş., Banvit, Şeker Piliç, Emek.
	OHSAS 18001	90,000–200,000 units/day	CP Gıda A.Ş, Keskinoğlu A.Ş.
	ISO 9001:2000	50,000–240,000 units/day	CP Gıda A.Ş., Banvit, Beypiliç, Şen Piliç, Keskinoğlu, Şeker Piliç, Köy-Tur, Emre Piliç, Lezita.
Fruitjuice	BRC-IFS	30,000 t/year	Tunay A.Ş.
	ISO 22000	25,500–250,000 t/year	Yasar Holding (Pınar), Asya Gıda, Yörsan A.Ş., Kızıklı A.Ş.
	HACCP-ISO 13001	30,000–300,000 t/year	Yaşar Holding, Dimes, Gülsan, Asya, Frigo-Pak, Etap, Yummy A.Ş., Yimpaş A.Ş., Tunay A.Ş., Meykon,
	ISO 14001	154,000–350,00 t/year	Yaşar Holding, Dimes, Yörsan
	ISO 17025	350,000 t/year	Yörsan A.Ş.
	ISO 9001:2000	200–350,000 t/year	Akman A.Ş., Aroma, Göknur, Gülsan, Asya, Etap, TAT, Yörsan, Yimpaş, Yummy, Kızıklı, Tunay, Meykon
Milk and milk products	BRC-IFS	2,100 t/day	Ak Gıda A.Ş.
	ISO 22000	250–450bin t/year	Bahçıvan A.Ş., Cebel A.Ş., Danone Tikveşli A.Ş., Eker A.Ş., Pınar A.Ş., Yörsan A.Ş.
	HACCP-ISO 13001	Daily 50–600 t (annually 120,000–150,000 t)	Dimes A.Ş., Ekiciler, Enka, İtimat, Kaanlar, Kars-karper A.Ş., Sütaş, Teksüt, Yörük, Yörükoğlu
	ISO 14001	120,000–350,000 t/year	Dimes, Pınar, Yörsan
	ISO 9001:2000	150–2, 100 t/day	Ak gıda, Aysüt, Bahçıvan, Eker, Enka, Güneysüt, İtimat, Kaanlar, Kars-Karper, Pınar, Sütaş, Yörsan, Yörük, Yörükoğlu

(continued)

TABLE 3.2. (continued)

Sub-sector	Certificate types	Capacity	Firms
	ISO 17025	1,200 t/day	Yörsan A.Ş.
	ISO 15161	1,200 t/day	Sütaş
Olive and oliveoils	USDA Organic	75 t/day, 24,000 t/year	EKİZ
	BRC	130,000 t/year	Zade
	ISO 22000	130,000 t/year	Zade
	HACCP-ISO 13001	75 t/day–130,000 t/year	TARİŞ, Ekiz, Zade, Oro-Altınç, Heybe
	ISO 14001	130,000 t/year	Zade, Komili
	ISO 17025	130,000 t/year	Zade
	ISO 9001:2000	24 bin–130,000 t/year	Tariş, Oruçoğlu, Ekiz, Zade, Oro-Altınç A.Ş.
Biscuit, chocolate and candies	IFS	1,350 t/day,25,000m²	Ülker
	BRC	1,350 t/day	Ülker, Halk
	HACCP-ISO 13001	100–1,350 t/day	Ülker, Anı, Şimşek, Eti, Halk, Saray
	ISO 14001	600 t/day	Halk, Saray
	ISO 18001	600 t/day	Halk
	ISO 9001:2000	100–600 t/day	Saray, Eti, Halk, Hazal, Azra, Anı
Instant soap	IFS, BRC	45bin t/year	Halk, Tukaş
	HACCP-ISO 13001	45bin t/year	Tukaş, Halk, Tamek, Aroset
Tomato paste	IFS	138–250bin t/year	Tukaş, Tat
	BRC	38bin–139,000 t/year	Tukaş, Merko
	HACCP-ISO 13001	3bin–139,000 t/year	Akfa-Akpa, Assan, Burcu, Demko, Merko, Tukaş
	ISO 17025	250,000 t/year	Tat
	ISO 9001:2000	3bin–250,000 t/year	Akfa-Akpa, Assan, Akson, Oraklar, Burcu, Demko, Merko, Tamek, Baktat, Tat, Tukaş
Pasta	ISO 22000	250–300 t/day	Golda, Besler
	HACCP- ISO 13001	90–300 t/day	Besler, Tat, Öğün, Berrak
	ISO 9001:2001	90–300 t/day	Oba, Nuh'un Ankara, Golda, Besler, Beslen, Pastavilla, Selva, Doğa, Piyale, Tat, Türkmen, Yayla, Öğün, Berrak

2. Legislation for food quality systems in Turkey

2.1. MINISTRY OF AGRICULTURE AND RURAL AFFAIRS

The General Code of Health Protection, adapted in 1930 from respective Swiss law, covered regulations regarding official food controls in addition to many other aspects of public health issues. Later in 1952 a new but this time more specific regulation, called Food Commodities Regulation was developed to cover further detailed aspects of food controls to be carried out by governmental bodies. Together with the individual food commodity standards developed by TSI (Turkish Standards Institute, a member of ISO) following its foundation in 1960 and some of which were mandated in the local market, these three documents formed the basis of official food controls in Turkey (Koc 2007).

The first comprehensive food decree-law in Turkey was the Decree No. KHK/560 which covers the production, consumption and inspection of foodstuffs enacted in 1995 which aims to protect public health against all possible food related diseases and all stages of food production are subject to inspection (Alpay et al. 2001). Then, a set of regulations were prepared and published in official gazette with authorisation law 560, specifically prepared for harmonisation of the national food control system with EU food laws. The pioneering ones were the regulation describing in detail the good manufacturing practices (GMP's) in food producing plants in 1996, the Turkish Food Codex which contains chapters similar to the horizontal EU legislation with corresponding specific communiqués on food additives, food contaminants, food packaging, food labelling and food hygiene. The Codex had foreseen the preparation of vertical codex documents covering individual commodity types, and as of today, many of these (including flour and bread, infant formula, fruit juices and nectars, alcoholic beverages, meat products) have already been prepared and replaced the previously mandated Turkish standards on these subjects. In addition, Turkish Food Codex includes the communiqués on nutrition labelling of foodstuffs and the addition of substances for specific nutritional purposes which have not been officially released, yet. After the release of EU food law 178/2002 (Regulation (EC) N° 178/2002)), the food decree-law 1995-KHK/560 modified and converted into "food law" May 27, 2004 (Law No: 5179 and published official gazette as of 5 June 2004, No: 25483). With this new food law, MARA has become the competent authority for inspecting all stage of food from production to consumption and took over all the responsibility for inspection of food safety. However, there was a great difference between EU food law and new Turkish food law. The last Turkish food law did not include feed and veterinary concepts, which must be intended to realize food safety controls from farm to fork.

Moreover, a draft law, including feed and veterinary issues, has been prepared to harmonise food law with EU food laws.

The increase of value-added products in agriculture sector is mainly attributable to crop production and changes in yield directly affect growth of the sector. In this context, in order to improve productivity and quality in plant production, it is aimed to improve quality of crops by supplying all types of reproduction materials related to the crops with high quality and higher genetic potential in line with standards, on time and with affordable prices. Therefore, the new Seed Law No: 5553, which was prepared to amend the Law on Registration, Control, and Certification of the Seeds (No: 308 and dated 1963) in line with the EU harmonisation process and by taking into account international seed systems and advanced technologies, was enacted in 31 October 2006.

Parallel to the global developments, for preservation of the ecologic balance, decreasing the negative impacts of agriculture on the environment and improving socio-economic level of producers, controlled greenhouse production, good agricultural practices and organic agriculture practices have become more important in Turkey.

With the regulation entitled "Regulation for Controlled Greenhouse Production (dated as of 27 December 2003 official gazette no: 25329)", Turkey has also started adapting the GAP standards related to important Turkish export products such as off-season greenhouse vegetables. Moreover, GlobalGAP (formerly used as EurepGAP) prepared by European supermarket chains and their suppliers, was converted into a Turkish legislation and published in official gazette no: 25577 "Regulation on Good Agricultural Practices" on 8 September 2004 (Regulation 2004). The requirement of Turkish legislation and GlobalGAP are almost the same, however, Turkish certification do not have a worldwide acceptance. Furthermore, different certification is also required such as Tesco's Nature Choice (TNC) by multinational supermarket chains.

The first By-Law of Organic Agriculture was issued in 1994, adopting the EU definition of organic agriculture following the entry into force of EC Council Regulation no: 2092/91 in 1991 (Regulation (EC) No 2092/1991)). The amendment was made to overcome some malfunctions of By-Law and the sanctions were added against the fault and inaccuracy in organic agriculture activities with the Regulation 22328 dated 29 June 1995 in official gazette.

In August 1999 rules on production, labelling and inspection of the most relevant animal species were agreed (Council Regulation (EC) No 1804/1999 of 19 July 1999), covering issues as foodstuffs, disease prevention and veterinary treatments, animal welfare, husbandry practices and the management of manure. In March 2000 the European Commission introduced with

Commission Regulation (EC) No 331/2000 of 17 December 1999 a logo bearing the words "Organic Farming-EC Control System". This logo can be used on a voluntary basis by producers whose systems and products have been found to satisfy Council Regulation (EEC) No 2092/91. Therefore, another regulation came into force in 2002 "Regulation on Organic Agriculture Principles and Applications" official gazette no: 24812 dated 11 July 2002. At the end, a comprehensive Organic Farming Law (No: 5262) has been in force as from December 1, 2004 is in line with EU Regulation 2092/91 (Organic Farming Law 2004). Moreover, new regulation under the new law came into force on 10 June 2005 in official gazette (No: 25841), but Turkish Organic Agriculture Legislation further needs to be improved to comply with new EU legislation.

According to data published in 2007, approximately 200,000 ha areas are used by 14,000 farmers for organic agriculture in Turkey. The majority of organic production is sold in foreign markets, primarily in European countries, and exports have been steadily growing. The domestic demand has started to increase since late 1990s and but still very small relative to total food demand. The legislation assigned MARA the responsibility of overseeing the cultivation of organic crops.

The Ministry created a specific Organic Agriculture Committee (OAC), which is the main decision-making body. It prepares and implements the regulation, authorises certification bodies, inspects these organisations and coordinates all other activities to improve and foster organic agriculture (Ozkan 2002). The OAC is composed of representatives of the various Directorates of MARA. In support to the work of the Ministry, the Turkish Association of Organic Agricultural, Wheat Association for Ecological Living and Organic Product Producers and Industrialists Association, non-governmental organisations, provides policy input, contributes to technical improvement and improve domestic market in the organic industry. The members of these associations include producers, exporters, academics and consumers.

With regard to the management infrastructure, Organic Farming Units have been established at the Provincial Agricultural Directorates. The aim is to provide help to certification bodies with specialised staff to collect data required by standards and perform inspections and certifications of companies as technical auditors. There are currently 13 certification bodies established in Turkey.

The exportation of organic products to EU is mandated for third countries by EC Council Regulation no: 2092/91 in detail as some rules must be put into their own legislations and they must apply to the EU with documents including the legislation and various technical and administrative issues for organic production. Turkey has a legal application in announced period through Turkish Ministry of Foreign Affairs to present in the list "Third Countries Exporting Organic Product to EU".

The preparation of legislation about trans-genetic organisms has been sustained parallel to the international legislations. MARA has started to prepare trans-genetic organism legislation after the meeting with research institutes, relevant public general directorates and universities on April 1998. This concept was divided into three parts as "Cultivation Examination of Trans-Genetic Crops", "Registration of Trans-Genetic Materials" and "Production, Marketing of Genetically Modified Organisms (GMO) and Usage as Foodstuff". However, only a directive, containing cultivation examination and registration subjects, was released on 14 May 1998 which was revised in 1999 to prevent conflicts with the "Regulation on Registration of Crops Varieties" (SPO 2000).

According to these improvements draft legislation has been prepared on "Production, Marketing of Genetically Modified Organisms (GMO) and Usage as Foodstuff". Thus, there is no legislation on specific controls of products containing GMO and no permission is given in case of any existence of GMO is declared in the product. Therefore, production and imports of Genetically Modified Organisms (GMOs) are not permitted or authorized in Turkey. However, General Directorate of Protection and Control (GDPC) of MARA is the competent authority for the control and MARA has four laboratories to analyse GMOs (MARA 2007). MARA indicates that the legislation on GMOs, novel foods and National Bio security Law have been prepared since 2000.

Animal species, increase production of concentrated feed and fodder crops of high quality, organise animal breeders, eradicate animal diseases and pests, and diversify publications are being improved in the livestock sector. The Decree No. 2005/8503 of the Council of Ministers on Supporting Animal Husbandry aims at increasing of production of roughage, promotion of breeding of studs, spreading of artificial insemination, and creating regions free of animal diseases (SPO 2007). In addition, works on the alignment of national legislation on livestock with the related EU regulations are underway. In this context, an animal identity system is almost completed including all livestock population in the system. Starting from 2005, works, regarding the operation and healthy functioning of the system, have been continued through the inclusion of new born into the system and removal of those animals, which died or slaughtered from the system. Works were commenced to identify sheep and goat population.

2.2. TURKISH STANDARDS INSTITUTION

Turkish Standards Institution (TSI) is an autonomous semi-public organisation, but according to law 132, promulgated in 1960, only those standards adopted by TSI shall be called "Turkish standards". Furthermore, the authorities have entrusted TSI with representing Turkey within the regional and international organisations dealing with standardisation. TSI is very active at the national level but also at the international level. In 2002 there were 18,129 Turkish standards in TSI catalogue and surpassed 20,000 in 2005.

2.3. TURKISH PATENT INSTITUTE

In 1992, the EU created systems known as PDO (Protected Designation of Origin), PGI (Protected Geographical Indication) and TSG (Traditional Speciality Guaranteed) to promote and protect valuable food names under Regulations (EEC) No 2081/92 and (EEC) No 2082/92.

PDO means the name of a region, a specific place or, in exceptional cases, a country, used to describe an agricultural product or a foodstuff. This foodstuff must be originating in that region, specific place or country and possessing quality or characteristics which are essentially or exclusively due to a particular geographical environment with its inherent natural and human factors and the production, processing and preparation of which take place in the defined geographical area.

PGI means also the name of a region, a specific place or, in exceptional cases, a country, used to describe an agricultural product or a foodstuff. This foodstuff, differently, must be originating in that region, specific place or country and which possesses a specific quality or reputation or other characteristics attributable to that geographical origin and the production and/or processing and/or preparation of which take place in the defined geographical area.

TSG is used for products with distinctive features which either have traditional ingredients or are made using traditional methods (Boel 2007).

The purpose of this regulation is to provide information on the inspection bodies notified by the Member States for each geographical indication or designation of origin registered. In March 2005, the World Trade Organization (WTO) released the panel report regarding the European GI system. The conclusions and recommendations of the panel led the European Union to revise its rules governing how international GIs are treated. Specifically, European Council (EC) Regulation 2081/92 was amended with Regulation (EC) 510/2006 (EC 1992, 2006; WTO 2005). The amendment is aimed at complying with the Agreement on Trade-Related Aspects of Intellectual Property Rights (TRIPS) of the WTO. In particular, the new regulation allows the EU regulatory system to recognize and protect foreign GIs and allows foreign producers to apply directly for registration of GI products in the European Union. According to regulation 510/2006, to obtain geographical indication certification, the applicant must apply to its national authority. The national applications are executed to publication and objection phase in their own country. After the last evaluation, the applications are transmitted to EU commission. In this concept, EU certification for Turkish products could be realized. However, there are some differences from Turkish Decree-Law No: 555 pertaining to the Protection of Geographical Sign as from 27 June 1995 and its regulation.

Regulation 510/2006 includes more detailed judgement from national legislation about the inspection of geographical signs such as inspection has to be executed by independent and specialized control bodies accredited with the standards EN 45011 or ISO/IEC Guide 65. This obligation must be accepted by both EU and non-EU countries before 1 May 2010. Therefore, draft legislations have been prepared to become in accordance with the new EU law and regulation EC 1898/2006 (Regulation (EC) No 1898/2006).

Protection of Geographical Indication is in force by "Decree-Law No: 555 pertaining to the Protection of Geographical Indication as of June 27, 1995" that covers both food and non-food materials although it was revised in November 7, 1995 (Decree-Law on the Protection of Geographical Signs 1995). There is also not any "Traditional Speciality Guaranteed" concept either in this decree-law or in any other regulation as EC Regulation 509/2006. However, last draft includes both protection of geographical signs and traditional speciality. The differences between EU and Turkish Geographical Protection Legislations are given in Table 3.3.

2.4. THE UNDERSECRETARIAT FOR FOREIGN TRADE

There exist a total of 52 inspection units called as "Inspectorates for Standardization for Foreign Trade", within the eight Regional Directorates (Marmara, Western Anatolia, South Anatolia, Eastern Black Sea, Western Black Sea, South Eastern Anatolia, Central Anatolia and Eastern Anatolia) under The Undersecretariat for Foreign Trade (UFT), General Directorate for Standardization for Foreign Trade. These inspection units carry out the issuance of the "Inspection Certificate(s)" only for some agricultural products to be exported/imported within the scope of product standards of Turkish Standards Institution (TSI).

In exports, the agricultural products within the scope of the 70 standards are subject to this kind of conformity inspection. Formerly, this inspection was only realised in the export phase whereas the same procedure is to be realised in the import phase as well according to this Regime. The inspection of the agricultural products within the scope of 70 standards prior to export and import is going to be performed by the Inspectors for Standardization for Foreign Trade. These inspections are performed according to the Turkish standards which are analogous to the respective OECD and EC standards. The exporters/importers should obtain Inspection Certificate from the "Inspectorates for Standardization for Foreign Trade" (UFT Technical Lepislation Law No. 4703, 2001).

TABLE 3.3. Differences between EU and Turkish geographical protection legislations.

EU regulations	Basic rules	Turkish laws	Basic differences
Regulation (EEC) No 2081/92 of 14 July 1992 on the Protection of Geographical Indications and Designations of Origin for Agricultural Products and Foodstuffs	– Lays down the rules on the protection of designations of origin and geographical indications for agricultural products intended for human consumption (except wine) – Only a group or a natural or legal person subject to certain conditions shall be entitled to apply for registration	Decree-Law No:555 Pertaining to the Protection of Geographical Signs in Force as from 27 June 1995	– Covers protecting the natural, agricultural, mining and industrial products and handicrafts – A group and natural or legal persons, who are producers of the product, could apply for protection
EC Regulation 510/2006 of 20 March 2006 on the Protection of Geographical Indications and Designations of Origin for Agricultural Products and Foodstuffs	– Only a group shall be entitled to apply for registration – Registration application concerns a geographical area situated in a third country could be achieved with the proof that the name in question is protected in its country of origin – A member state or a third country may object to the proposed registration	Draft Law – law on pertaining to the protection of geographical signs and traditional speciality guaranteed products	– Includes both geographical signs and traditional speciality. – Only a group could apply for protection – Regulates the relationship with the international protection, such as protection in EU

On the other hand, industrial products within the scope of 614 product standards, which are at the same time mandated in the domestic market, are subject to inspection by the TSI. The importers should also obtain Conformity Certificate from TSI (similar to CE mark in EC) before the importations, since the products should be in conformity with the relevant standard or regulation or technical document in respect of minimum requirements of human health and safety, animal or plant life or health and protection of environment, thus providing adequate information to the consumers at the stage of actual import.

In the import of the industrial products, which are in the scope of the Turkish Standards mandated, if the importer declares that the product is in conformity with the relevant international standards (ISO, CEN, IEC, CENELEC, ETSI), the inspection may be realised, upon request, due to these international standards. For the products which are already certified according to regulations of the European Communities ("CE" Mark, "E" Mark, "e" Mark, etc.) and freely circulated in the European Union, a Conformity Certificate shall be issued directly in case that the technical file submitted to Turkish Standards Institution before the import stage.

The Regime also regulates the controls to be carried out by Ministry of Health, Ministry of Agriculture and Rural Affairs and Ministry of Environment and Forestry. Pursuant to the Communiqué No.2005/5 which is related to human health and safety and Communiqué No.99/5 which is related to human health, animal and plant life and health, the importation of certain specific goods is subject to Control Certificate by Ministry of Health or Ministry of Agriculture and Rural Affairs. The scope of these product groups are as follows:

- Pharmaceutical products, drugs, some consumable medical products, cosmetics and detergents (Ministry of Health)
- Foodstuffs, agricultural and animal products, veterinary products and products used for agricultural protection (Ministry of Agriculture and Rural Affairs)

In order to fulfil obligations arousing from the Customs Union with European Community, products within the scope of the Communiqué dated 14 February 2004 and numbered 25373 and amended in the Turkish official gazette dated 17 November 2005 and numbered 25996 which are bearing a CE Mark and freely circulated in the European Union, a Conformity Certificate shall be issued directly in case that the technical file submitted to the Ministry of Health before the import stage.

In order to obtain a Control Certificate, pro-forma invoice of the said product(s) must be presented to the related ministry. Furthermore, depending on the type of the product, the following documents shall be presented to the ministry: health certificate, certificate of analysis, formula or list of contents of the product, pedigree certificate, radiation analysis report, etc.

The above mentioned documents, particularly health and/or analysis certificate should be obtained from and/or approved by the public authorities of the producer country. These documents should be in original and translation is required for each document, including the pro-forma invoice. Control Certificate(s) must be obtained prior to import and presented to Customs Administration during actual import stage. Validity of the Control Certificate(s) changes from 6 to 12 months, depending on the inspected product.

3. Applications within the food quality system

According to the Organic Agriculture Law and Regulation, organic fresh vegetable and fruits are exempted from wholesale market law[4] that make compulsory for wholesale or retail sale to pass through the wholesale market halls. According to this law 14.4% deduction is applied to the producer's price including hall agent's fee as 8% of the selling price.

In Turkey, basic Direct Income Support (DIS) has been applied to every land on which vegetal production is made and additional payment has been made out of DIS to the farmers who get soil analysis done and perform organic farming since 2005. It was stated that the maximum amount of additional DIS that would be given to the farmers dealing with organic farming would be as much as the amount that forms the basis for DIS.[5]

Subsidized credit has been implicated since 2004 to encourage farmers producing organic, ITU and Controlled Covered Greenhouse production.

Agricultural support for controlled greenhouse production and ITU (GAP) is only subsidized credits. 40% deduction is applied to 17.5% nominal interest rate of Agricultural Bank for the farmers implying controlled greenhouse production. Furthermore, 60% deduction is applied to nominal interest rate for the farmers implying organic farming or ITU (GAP) certificated. Moreover, 50% support for usage and production of certificated seed is provided to registered farmers of Agricultural Registration System by MARA.

---

[4] The exemption was published in Organic Food Law (No: 5262) dated December 3, 2004.
[5] DIS is 10.00 TL/da in 2008. Additional payments soil analysis and organic farming are 1.00 TL/da and 3.00 TL/da, respectively. 1.70 TL = US$1.00

4. Summary on food defense

Food terrorism is defined by the WHO (2002) as: "an act of threat of deliberate contamination of food for human consumption with chemical, biological or radio nuclear agents for the purpose of causing injury or death to civilian populations and/or disruption of social, economic or political stability". In this context "food" includes crops, farm animals, minimally processed and processed foods and water (whether for drinking, use as a food ingredient or for use in food processing). By extension, so-called "eco-terrorism" covers the ideologically-motivated destruction of crops or animals and associated research facilities.

All societies are crucially dependent upon the food supply; therefore, its disruption is an obvious prime target for terrorism. Although explosive devices have been the favourite tool of environmental, animal rights and political terrorists to date, a number of different materials have been used to contaminate consumer goods, foods, and drugs across the world. The deliberate introduction of plant or animal diseases could also cause widespread disruption of the food supply.

4.1. RESPONDING TO ACTS OF TERRORISM AGAINST THE FOOD SUPPLY

Responses to terrorist acts may be divided into three categories. The first, is directed towards the immediate treatment of those affected; the second, involves a criminal investigation and apprehension of the perpetrators; and the third, minimizing casualties through the recall of the suspect product(s); the detection of the causative agent or its vector, and limiting the spread of the contamination. Governmental and private sector planning activities must focus on response measures and the development of preventive measures to protect product, facilities and members of the community as well as traceability measures for tracking and recovering affected materials.

4.2. EFFECTIVE FOOD CONTAMINANTS

Biological materials and chemicals are the most likely agents for food contamination. There are several toxins could be easily dispersed into a food, would survive a conventional thermal process used in food processing, and are stable under acidic conditions. Many of these are difficult to isolate and detect in a complex food matrix or take a long time to recover and identify. An agent that would impart little change to the sensory properties of a food so neither food sellers nor consumers would be suspicious that the food had been contaminated would pose the greatest risk. The most effective ones would be potent and easy to conceal. Despite governmental efforts to study more exotic materials, the most likely agents for a food contamination event remain common industrial chemicals and microbes with which the food industry and public health professionals are familiar.

To be effective, a small amount of contaminated product should be sufficient to harm large populations and/or cause injury or damage over a broad geographic region; this is potentially the greatest risk to agriculture through introduction of a crop or animal disease. The contagious nature of some disease causing microorganisms makes it possible for one infected individual to continue to spread the disease to others depending upon the agent used. One of the most worrisome food defense scenarios is a surreptitious attack with an agent that produces symptoms that are easy to misdiagnose. In this situation, the first responders are likely to be health professionals rather than law enforcement or other traditional first responders. Under such a scenario, a terrorist-induced epidemic could go unrecognized and undiagnosed for a significant period of time, delaying treatment and other control efforts for containment and quarantine. The effectiveness of a possible agent can be based on the following factors:

- Potential impact to human, animal or plant health
- The type of food material contaminated
- Ease of detecting contamination of the food through discernible changes in appearance, odour or flavour
- The point in the food supply chain where the contamination was introduced
- The potential for widespread contamination
- The fear people would have from the use of the contaminant or the particular food to spread illness or disease

4.3. FOOD MATERIALS AND PRODUCTS AT GREATEST RISK

A wide variety of food products are at risk including those that are perishable, ready-to-eat, and frequently consumed. They have a rapid turn around time and which would be consumed before detecting the hazard. Also at high risk are foods or food ingredients prepared in large batches into which a toxic agent could be dispersed throughout a large quantity of material (including via water), and then into numerous servings of a wide variety of products. Because food is distributed rapidly, often over great distances, and to large populations in different locations, this creates a potential for widespread impact. The efficiency of food distribution could make it difficult to mount an appropriate response because the public health impact would occur within a matter of hours or days. The systems in place for producing and distributing food also create opportunities for intentional contamination (USDA 2003). Food and agriculture products are commonly accessible at some point during growing, harvesting, processing, storage and distribution. The unit operations involved in food production such as mixing, incorporation of minor components, dilution, or size reduction could spread a toxic material throughout a large batch or into numerous batches of product. Spreading contamination throughout a facility also poses a food safety as well as an occupational hazard and would result in a facility being taken out of service until it could be decontaminated.

4.4. ADDRESSING A HOAX

Accessibility to a specific type of attack, the method of the attack, and controls in place to deter such an attack, are all critical to determine if an incident is a hoax. Here a vulnerability assessment in conjunction with a security analysis can be used to determine exposure and evaluate weaknesses within a business operation. Questions to ask are:

- Could the attacker reach the product without detection?
- Could an attacker bring enough material through the conventional security infrastructure at the facility?
- Could an attacker introduce enough material at a point in the process to successfully contaminate the product at levels high enough to cause harm without this action being detected?
- Could the contaminant be detectable in the product by a simple testing?

Having appropriate detection and mitigation strategies in place are important. A company must be able to tell law enforcement, government regulators, and the media, that enough barriers were established within the operation to deter the alleged attack and that these barriers could not have been breached without detection. In the case of a hoax, a company must be able to say, with certainty, that the contamination did not happen, making a statement that it is likely that it did not happen is not good enough anymore (Ryan 2005).

4.5. MITIGATION STRATEGIES

The basic tenets of threat assessment take into consideration the value of the asset to be attacked, the vulnerability of the asset, the likelihood of an attack and the intent and capability of the attacker. Food businesses face the task of reducing the capability and likelihood of an attack by addressing foreseeable risks and developing effective and efficient means to reduce the risk and can be implemented within a facility. One workable and simplified food defense awareness program is called ALERT. This pneumonic identifies the key five food defense points and stands for: Assure Look Employees Report and Threat (IUFoST 2007).

5. Conclusion

Food and agriculture are part of the critical infrastructure in every country in the world and it behaves everyone involved in these sectors to support the establishment of a coordinated strategy at the regional and national level to protect the food supply by conducting reasonable risk assessments and developing realistic defensive strategies (Goodman 2006) to address the risk of intentional contamination. Food has been targeted in the past, and is a relatively soft target. Recent developments in food defense include more sophisicated models for risk assessment with the current focus on devising protective measures to ensure a safe food supply. As part of this, the food industry needs to have the support of governments at all levels to support the implementation of realistic and workable food defense programs without the imposition of additional regulatory burden, cost, or impediments to trade. The role of governmental entities remains to ensure that response efforts are rapid, and coordinated in such as way to protect public health, and that avoid confusion and unnecessary duplication of effort. It remains our hope that large-scale intentional contamination incident involving the food supply will not occur, but if this should happen that we shall be prepared to respond and limit its impact.

The information flow from bottom to top in a time of crisis does work properly in the light of the experience(s) of EU-member states with "Rapid Alert system for Food and Feed (RASFF)". The traditional physical security practices alone cannot protect the food sector as agriculture and food systems are extensive, open, interconnected, diverse, and complex structures providing attractive potential targets for terrorist attacks. Due to the rapidity by which food products move in commerce to consumers and the time required for detection and identification of a causative agent, attacks on the food and agriculture sector-such as animal or plant disease introduction or food contamination-could result in severe animal, plant, or public health and economic consequences. It is also a fact that a protection plan for food and agriculture infrastructure and resources must focus on planning and preparedness, as well as early awareness of an attack. Science-based surveillance measures are essential for recognizing a possible attack on the sector so that rapid response and recovery efforts could be implemented to mitigate the impact of an attack. A protection plan must also be coordinated closely with response and recovery plans.

Changing trends in food safety and food chain security should be taken into consideration. The amount of foods consumed outside the home is increasing throughout the world and these changes the trends in food safety. Justification for the concerns with food chain security is a fact; in particular the awareness that manuals for intentional contamination of food are widely available, that the use of biological or chemical weapons against the food supply could cause mass casualties and that even an ineffective attack could cause significant economic and psychological damage.

The success of communication is very important on food terrorism and also for possible use of food as a vehicle for terrorism, the public confidence to government(s) and-or authorities. Case studies should be detailed on how the surveillance systems, legislation and recall procedures function effectively to contain an incident or breach in food chain security by limiting the number of affected individuals. The risk communication should be centred on the consumer perception of risk both with regard to deliberate (terrorist) threats and to accidental or inherent food hazards. The number of possible barriers should be identified which affect the way in which a consumer will respond to a particular message. The matter of trust in the organisation issuing the information is of particular importance. A number of components could be identified that impact or contribute to this trust and the significant cultural differences should be shown to exist in relation to them. Differences should also be identified in the way people use risk information, further complicating the issue of risk communication.

The role of the media and how it could be used by a terrorist organisation is also important. In this respect the issues surrounding the relationship between the media and government organisations trying to address (risk communication) should be considered as a food safety issue. The real consumer perception of food chain security hazards has to be taken into consideration. Finally, the issues related with Food Defense will continue to be a priority topic for all countries as it is a global issue that is continually developing.

Acknowledgements

The first three parts of this chapter was quoted from the initial report of Food Quality Assurance Schemes in the Candidate Countries: Turkey (FQAS-TURK) Project executed by the authors. The project was funded by EC Joint Research Centre – The Institutes for Prospective Technological Studies in Seville (Contract Number: 151089-2008 A08-TK, 2008).

Bibliography

Alpay, S., Yalcin, I. and Dolekoglu, T., 2001, Export Performance of Firms in Developing Countries and Food Quality and Safety Standards in Developed Countries (mimeo) (1 September 2004) (http://www.econturk.org/Turkisheconomy/Exportperformance-Alpay Yacin&Dolekoglu.pdf)

Boel FM (2007) European Policy for Quality Agricultural Products. EU Fact Sheet, Germany Decree-Law on the Protection of Geographical Signs, Law No: 555 Official Gazette published : 27th June 1995/22326

FAO (1996) Decleration on world food security. World Food Summit, FAO, Rome

Goodman, T. 2006. Connecting food safety and food security. Presentation to the Association of Food and Drug Officials. Public Health Administration and Food Security Specialist. Division of Food Protection, Indiana State Department of Health.

Grunert, K.G., 2002, "Current Issues in the Understanding of Consumer Food Choice", Trends in Food Science and Technology, 13 (8): 275–285

Guittard, C., 2006, Food Safety in Turkey (IP/A/ENVI/OF/2006-104), DG Internal Policies of the Union, Economic and Scientific Policy Department. (http://www.europarl.europa.eu/comparl/envi/pdf/externalexpertise/ieep_6leg/food_safety_in_turkey_guittard.pdf)

IUFoST, 2007. Scientific Information Bulletin September 2007.

Koc A. A., 2007, The Food Quality Management System in Turkey (chapter within the report, in Italy) Published by the ISMEA-IAMB, Italy

Law on Adoption of The Amended Decree By-Law on The Production, Consumption and Inspection of Food Law No: 5179 Official Gazette published: 5th June 2004/25483

Oskam A., Burrel, A., Temel, T., Berkum, S., Longworth, N., Vilches, I.M., (2004), Turkey in the European Union, Consequences for Agriculture, Food and Rural Areas and Structural Policy, Final Report, Wageningen University (http://www.econturk.org/Turkisheconomy/turkey-eu-agriculture.pdf)

Ozkan, M. 2002. Organic Agriculture and national legislation in Turkey. In Organic agriculture: Sustainability, markets and policies. OECD workshop on organic agriculture, Washington, D.C., USA, 23–26 September 2002, pp. 289–294 (http://www1.oecd.org/publications)

Pinstrup-Andersen, P. 2009. Food Security: definition and measurement, Food Security 1: 5–7

State Planning Organisation (SPO), 2007, Food Safety, Sanitary and Phytosanitary Special Commission Report (OIK), 9th Development Plan (2007–2013), Publication Number: SPO. 2711, OIK: 664 (http://ekutup.dpt.gov.tr/oik/plan9.asp, access date August 2008)

State Planning Organisation (SPO), 2000. Biotechnology and Biosecurity Special Commission Report (OIK), 8th Development Plan (2000–2006), Publication Number: SPO. 2515, OIK: 533 (www.dpt.gov.tr/DocObjects/Download/3026/oik533.pdf, access date March 2009)

Regulation (EC) N° 178/2002 (28 January 2002), European Parliament and the Council, General principles and requirements of food law, establishing the European Food Safety Authority and laying down procedures in matters of food safety

Regulation (EC) No 2092/1991 (24 June 1991), Regulation on organic production of agricultural products and indications refers thereto on agricultural products and foodstuffs

Regulation (EC) 510/2006 (20 March 2006), Regulation on the protection of geographical indications and designations of origin for agricultural products and foodstuffs

Regulation (EC) No 1898/2006 (14 December 2006), Regulation laying down detailed rules of implementation of Regulation (EC) 510/2006 on the protection of geographical indications of origin for agricultural products and foodstuffs

Ryan, R. 2005. Risk assessment to drive research on contaminant detection. Proceedings of the Institute of Food Technologist's First Annual Food Protection and Defense Research Conference. Nov. 3–4, 2005 (www.ift.org)

Regulation, 2007, (9th December 2007), Inspection and Control of Food Safety and Quality, Official Gazette published: 26725

Regulation, 2004, (8th September 2004), Good Agricultural Practices Official Gazette published: 25577

Organic Farming Law, 2004, (1st December 2004), Law No: 5256 Official Gazette: Regulation on Essentials and Implementation of Organic Farming Official Gazette published: 10th July 2005/25841

The Law Relating to the Preparation and Implementation of the Technical Legislation on the Products, Law no: 4703, 11/07/2001-24459 (Undersecretariat for Foreign Trade (UFT)

TUIK 2009, Periodic Results of Household Labour Force Survey (Turkey, Urban, Rural) (http://wwwtuik.gov.tr, access date February 2009)

USDA (2003) FSIS Safety and Security Guidelines for the Transportation and Distribution of Meat, Poultry, and Egg Products. August, Washington DC

USDA 2004, Turkey Exporter Guide Annual 2004. Foreign Agricultural Service, GAIN Report TU#4008, 3/17/2004

WHO (2002) Food Safety Issues. Terrorist Threats to Food. Guidance for Establishing and Strengthening Prevention and Response Systems, World Health Organization, Food Safety Department, Geneva, Switzerland

World Bank 2006, Turkey Country Economic Memorandum Promoting Sustained Growth and Convergence with the European Union. Volume I: Main Report" Report No. 33549-TR. February 23, 2006. (http://siteresources.worldbank.org/INTTURKEY/Resources/361616-1141290311420/CEM2006_Main.pdf)

MONITORING OF ENVIRONMENTAL RESOURCES AGAINST INTENTIONAL THREATS

NELSON MARMIROLI, MARTA MARMIROLI, AND ELENA MAESTRI

Department of Environmental Sciences, University of Parma, Viale G.P. Usberti, 11A, 43100, Parma, Italy
e-mail: nelson.marmiroli@unipr.it; marta.marmiroli@unipr.it; elena.maestri@unipr.it

Abstract: Terrorism aims to attack resources which are critical and vulnerable, but it often aims also at jeopardising aspects of human life which have the highest emotional impact. One of these aspects can surely be the environmental resources, since everything dealing with safety and salubrity of the environment affects all strata of the population. In particular, when environmental contamination affects the production of food and the supply of water the impacts become higher: it affects populations at large, it specifically affect the weak members of the populations, e.g. children and elderly, it addresses a fundamental need of people, and its perturbation is of high psychological impact, its manipulation can destroy the consumers' trust in industry, producers, and retailers, and it leads to massive economical impacts. This paper describes instances of environmental contamination and lists possible sensitive targets for terrorism. It lists current and innovative approaches to monitoring of resources and describes the initiatives organised within the NATO Science for Peace and Security Program against ecoterrorism.

Keywords: Countermeasures, ecoterrorism, environmental resources, food safety, monitoring, NATO Science for Peace, risk characterisation, sensors

1. Introduction

History has given us many examples of environmental resource been used as warfare target. Early as in the sixth century B.C. Assyrians poisoned water wells with human and animal bodies to induce epidemics. It was reported that Native Americans were given blankets containing smallpox viruses by the British, targeting those tribes which were loyal to the French Army.

During the Second World War several strategies for biological warfare were actively considered by Germany, Russia, Great Britain, Japan, United States (Guillemin 2006). In more recent times, a well known example dates from 1978, when the Bulgarian dissident Georgi Markov was assassinated with ricin toxin injected via an umbrella tip, by agents supposedly sponsored by the Bulgarian government.

Important areas of vulnerability to terrorist threats in the modern societies are food and beverage supply chains, including drinking water, agriculture and farming, water supply, processing, transport and distribution to retailers and catering of foods to end-users, i.e. consumers.

A lack of a security policy for the food products, and lack of control at critical points of the supply chains, exposes the citizens to possible terrorist actions, with toxic biological and chemical agents.

This paper will consider environment as the main target for terrorism and its vulnerability, illustrating examples of environmental contamination. Then it will describe the main agents of concern and how they can be detected anticipating the damages they may cause. At the end, it will describe the NATO Science for Peace Program "SITCEN" and its mission for developing an International Situational Center as countermeasure against environmental terrorism.

2. Vulnerable environmental resources

Vulnerability of environmental resources is a fact actively exploited by terrorists. These resources can be used indirectly to carry on damages into human populations, or be the direct "targets" of an action (Chalecki 2002). For instance, water can be used as a vehicle to carry on pathogens to affect human populations, but through the disruption of the water supply infrastructures can determine water deprivation or water shortage. Criteria for establishing vulnerability are: the scarcity of a resource, its importance, its location and accessibility, the capacity for its regeneration. Additional critical factors can come from cultural or political considerations, because some targets could have an increased impact according to its social or cultural perception.

2.1. WATER

Water can be a potential terrorist target; either surface water or aquifers. For subsurface water, there are many examples of contamination, mainly coming from soil pollution and leaching of organic contaminants, as for industrial pollution. Surface water can be plagued with water-borne pathogens, chemical substances, viruses, bacteria, and protozoa.

Vulnerability of water as a resource is quite evident: it is essential to population and there is no surrogate for it. In military history many examples of contamination or targeting of water resources are reported (Anderson 2003). The Roman Emperor Nero used laurel water containing cyanide to poison wells of the enemies. General Grant destroyed river flow during the civil war siege to Port Hudson in 1863. General Chiang Kai-shek destroyed dams of the Yellow River to flood areas threatened by the Japanese army in 1938.

Concerning water reservoirs, they may be located far from the urban areas and surveillance in this case can be small. So the risk is that terrorists can act without being detected; in this case the resource has a high accessibility. On the other hand, it is also true that substantial amounts of chemicals or pathogens may be needed to pollute the water provision of a population.

However, examples of accidental outbreaks show how an intentional contamination could be successful: in April 1993 an outbreak due to *Cryptosporidium* in Milwaukee brought to the death of more than 100 people, and to disease in 400,000 people. This outbreak was presumably due to malfunctioning of the treatment plant and to illegal discharges: *Cryptosporidium* oocysts from Lake Michigan entered into the water system and were not removed during the filtration processes. Possible sources of contamination could have been cattle, slaughterhouses and human sewages (Mac Kenzie et al. 1994). As a consequence, a practice of monthly testing for this microorganisms came into force.

In July 2000, in France, the Meuse river was poisoned by red coloured sulfuric acid: the pollution was provoked by workers of a chemical plant protesting against their employer (Chalecki 2002). This has shown the vulnerability of water bodies towards intentional threats.

Concerning the accessibility of water infrastructures, the backflow in water pipes can be exploited to distribute toxic chemicals. Cross connections among pipes can cause instances of backflow. Also, an external system operating at higher pressure when connected can send contaminants into the distribution system, and from here to other pipes. Some examples of accidental backflow have occurred in instances of inappropriate connections of potable water and non-potable water sources (www.defendyourh2o.com/).

The failure of a wastewater treatment plant can cause major damages to the agricultural system and to the environment. Failure could be caused by excessive biological load or by substances damaging the biological processes, and of course by physical damage of the plant. In Italy only large treatment plants, dimensioned for over 10,000 equivalent inhabitants, are equipped with remote control systems that evaluate continuously the water quality.

During the meeting "ECOTER Workshop on Ecosecurity: Securing water resources against terrorism" held in Budapest (Hungary) on 5–6 October 2006, several speakers addressed the topic of water resource vulnerability.

Prof. Istvan Szabò (Szent István University, personal communication) showed how a Geographical Information System can be used to build maps of Hungary showing the distribution of highly sensitive areas for the status of subsurface waters, including for instance points of water supply for drinking purpose, and also areas protected by UNESCO. Different levels of sensitivity can be attributed to different water bodies and the maps can show their distribution in relation to military activities, or to oil transporting pipelines. This is of vital importance to identify critical points where terroris threats could be more harmful to environment and humans.

In that same meeting, Dr. H. Joel Allen (EPA, personal communication) showed how U.S. EPA is exploiting systems for monitoring water bodies with "sentinel organisms" sensing toxicity. The National Risk Management Research Laboratory has among its tasks the protection of watersheds, and monitoring is an essential component. Early warning systems (EWSs) use tools for monitoring water quality, and then can transmit data at a distance to a center which analyses them, distributing the information to decision makers. Early Warning Systems based on toxicological evaluation are not diagnostic, in fact they cannot identify the component responsible. Several tests are under development and validation, using living organisms: algae, bacteria, and shellfish. Biological changes in the species are taken as evidence of an alteration in water quality.

2.2. FORESTS AND AGRICULTURE

Rural areas in all countries are those mostly involved in food production, both for planta and for animals. Being at the basis of human nutrition, they experience a lack of infrastructures and may be associated to poorer comunities, as compared with metropolitan areas. Possible terrorist attacks can take advantage of the vulnerability of targets which are critical in food production (Stamm 2002).

Release of chemicals or radionuclear agents is a possibility, since manufacturing plants for chemicals, nuclear power plants and energy production facilities, but also storage sites for agricultural chemicals, are located out of urban centers. This may give the possibility for programming explosions or fires, which will cause an uncontrolled release of chemicals. Also, incidents around energy production plants can have indirect consequences on populations receiving and utilising the same energy.

According to Chalk (2004) the main impacts of agroterrorism would be the following: (i) economic disruption of critical infrastructures with direct losses, indirect costs and international effects; (ii) loss of trust in the government capacity; (iii) social instability due to fear and anxiety.

Biohazards are also a concrete threat: food of vegetal or animal origin as well as water resources are generated outside urban areas, and their contamination with pathogens or with toxins can be achieved with realtive efficiency with a devastating effect on the consumers in the cities (Pellerin

2000). Serious animal diseases amenable to terrorist exploitation are foot and mouth disease (FMD), avian influenza (AI), and Newcastle disease (ND). ND is a viral disease which can affect birds and poultry with lethal damages. AI also affects birds, it is viral and contagious. FMD is also viral and can be spread by animals and humans (Manning et al. 2005). Choosing an agent which is not pathogen to humans can be more amenable to the terrorist perpetrating the attack, because it can produce the most of the damage with the less of risk. Contagious diseases have the advantage that they will be transmitted to a large number of animals.

An updated list of transmissible diseases for animals can be found in the site of OIE, World Organisation for Animal Health (www.oie.int).

The recent example of AI in Asia and Europe showed how impacting can be such a disease, especially at the level of public opinion: only few cases have affected a large number of countries, evidencing a lack of experience and resources, manifesting a large economic impact on production and commercialisation of birds and poultry (European Commission 2006).

A possible target of agroterrorism can be rabbits for the haemorrhagic disease as occurred in New Zealand in 1997; local authorities believed that the disease had been spread deliberately in the attempt to control the rabbits populations (Carus 1998).

Forests are susceptible to fire or to herbicides action. Examples of deforestation through spread of herbicides have been part of our recent warfare history. Forests can present additional factors of vulnerability in their symbolic value: in some countries forests can even have cultural and religious importance. Commonly, forests are a source of income which will be depleted by the attack. The regeneration of the forest and therefore of the resources will be very slow. A disruption of economically important olive trees in Palestine occurred in 2000 as consequence of an act of war.

Examples of intentional threats are quite rare. In 1981 a group called "Dark Harvest" threatened to disseminate soil contaminated with anthrax in the United Kingdom. They wanted to call attention towards the dangers of biological warfare (Carus 1998). In Sri Lanka the tea plantations were threatened by the Tamil groups in 1986: possibilities were potassium cyanide used to poison the plants or foreign diseases to infect them (Carus 1998).

2.2.1. Food as a special vulnerable resource

A target for possible terrorist threat may be food contamination with chemicals or biological agents to provoke poisoning or infection. Examples of the disruptive potential of food contamination have been described. A widely known case is the dioxin contamination in meat as it occurred in Belgium in January 1999: 40–50 kg of oil containing PCBs, probably originating from waste recycling, were mixed to fats supplied to ten producers of animal feed. As a consequence, 500 t of contaminated animal feed were distributed to poultry farms and to other farming activities, even exporting to Netherlands, France and Germany, reaching out more than 1,800 farmers. In June all poultry and poultry-derived products were recalled from the market and destroyed if shown to contain more than 200 ng PCB per g of fat (van Larebeke et al. 2001). One important consequence of this episode was to bring to the introduction of food traceability in all steps of the food supply chain, whereas a direct consequence was hampering the poultry meat market for several months.

An important episode showing how the food chains are vulnerable to contaminants occurred in Spain in 1981: more than 800 people died and 20,000 were affected because of a chemical agent contained in a cooking oil.

Examples of contamination with microorganisms are also frequent. The best known episode is the outbreak of *Salmonella typhimurium* in the USA in 1985, which affected 170,000 people: the contamination occurred through pasteurised milk (Bean et al. 1990).

Food can be contaminated at many different points of the supply chain: production, transportation, processing, storage, final preparation and catering. An additional complication concerns the possibility that the contaminated food can be disseminated and transported covering long distances before the contamination was discovered. The food distribution may add additional risk factors, eventually.

The potential risk of terrorist attacks to food has to be taken seriously, because of the direct and indirect effects which can be induced. The direct effects as to the health of the population, with production of disease or of intoxication. The indirect effects are on the economic side consequent to either disruption of the food chain and weakening of consumer trust and confidence. These are the similar consequences as those observed in many instances of food contamination of accidental origin.

In 1989 some Chilean grapes happened to be poisoned with cyanide. It had no serious health effects but caused panic, and as a consequence, the Chile fruit export industry lost million of dollars. Inspections on food products reached a total of likely 8 million cases at major ports of entry. This examples showed how easily a "threat" of a terrorist attack was enough to damage a portion of the market. It also showed how controls when made at ports of entry could become effective as a countermeasure.

An accurate mapping of the supply chains and the identification of its critical points is basic to organise safety measures. It must be recognised that this is also a requirement for the HACCP (Hazard Analysis and Critical Control Points) procedure. Therefore, the activation of safety and precaution measures for food safety and quality can become beneficial also as a countermeasure against terrorism. When a HACCP and a traceability of the whole food chain are in place, the same is basically secured. But still, it would be very useful to have in place tools and strategies for monitoring at critical points for the presence of chemicals and pathogens. Controls and measurements of safety parameters, especially if they are established on-line or at-line along the suppy chain, would be an additional insurance (Marmiroli and Maestri 2007). Nowadays commercial instruments and diagnostic kits are available to check for the presence of microbiological hazards. However, these have mainly been designed to satisfy the requirements of legislators, such as in the Regulation 1441/2007 (European Commission 2007) for *Escherichia coli*, *Salmonella* spp., *Listeria monocytogenes*, Enterobacteriaceae, coagulase-positive staphylococci. Other tools are provided for additional organisms of particular interest for specific industries, like *Campylobacter* and *Clostridium*. Kits or tests can be more or less specific, targeted towards some strains, or species, or genera. Innovation and research is currently progressing towards the development of tests giving at the same time information on many types of different microorganisms. Multiplexed analysis of nucleic acids can offer a solution. A sample from a food matrix is analysed after DNA extraction, multiplex amplification with polymerase chain reaction, with a series of identification techniques based on specific probes to complementary targets (Marmiroli and Maestri 2007). But identification techniques can be based on more direct electrochemical reactions, in which successful recognition between the target organisms(s) and a receptor on a biosensor triggers an electric signal which can be transmitted, converted, measured and interpreted.

Complete safety is impossible to reach, because an assay cannot determine the presence of any possible biohazards in food products. Moreover, it would not be economically feasible to check for all. Safety relies on (i) effective HACCP procedure, to limit as much as possible the occurrence of attacks and (ii) traceability to allow for backtracing in case of accidents and perform effective recall of contaminated products.

An example cited by many papers on bioterrorism concerns the episode of food poisoning in The Dalles, Oregon, in 1984. A group is reported to have contaminated different salad bars with *Salmonella* and as a result 751 persons were affected. The motivation behind the attack was influencing a local election (Carus 1998).

Other examples seem less linked with terrorism and more to personal initiatives. For instance, in 1996 personnel of a hospital in Texas was infected with *Shigella dysenteriae* by eating pastries which had been contaminated by a former employee seeking revenge (Sobel et al. 2002).

Another motivation is economic damage: several instances are reported in which anti-Israel groups disseminated rumours concerning contamination of citrus fruit and grapefruit with liquid mercury and other substances. The result was hindering of exports from Israel to other countries (WHO 2008).

3. The main hazardous agents for intentional threats

We will discuss two main kinds of agents which have been used or can be used to perpetrate intentional attacks to environmental resources (Table 4.1). Whatever the agent, three main routes of exposure are possible for human beings: inhalation, dermal contact, ingestion.

3.1. CHEMICAL AGENTS

Toxic chemicals have been used as weapons since the World War I. They may be used to cause the death of people, or for disabling momentarily. Only few chemicals have the characteristics for terrorism purposes, and a complete list is reported by WHO (2004).

Sarin and VX are nerve agents; hydrogen cyanide is a blood gas; phosgene, chloropicrin, perfluoroisobutene (PFIB) are asphyxiant; mustard gas and lewisite are vesicants; lysergide and BZ (salt of 3-quinuclidinyl benzilate) are psychotropic; Adamsite, 2-chloroacetophenone (CN), 2-chlorobenzalmalononitrile (CS) and dibenz-(b,f)1:4-oxazepine (CR) are irritant. Most of these substances must be inhaled or are effective through direct contact with skin and eyes. Lysergide and BZ can be administered with food and drinks.

3.2. BIOLOGICAL AGENTS

Biological agents include living organisms or toxins produced by them, as opposed to manufactured chemicals. Production of organisms for terrorism can be accomplished with so-called "dual use" equipment, facilities which are of common use in several applications for food, microbiology, biotechnology but that can be used to grow hazardous organisms or extract toxins. An exhaustive list is reported by WHO (2004).

Even if several biological agents are of likely application, few are considered highly probable: anthrax (inhalation), smallpox, pneumonic plague and botulinum toxin.

Anthrax, *Bacillus anthracis*, is infective through spores. These may enter the human body via wounds in the skin, or by inhalation, or ingestion with food. Extensive heating destroys the spores, so the risk comes mainly from ready-to-eat food (Sobel et al. 2002).

Smallpox is caused by a virus and is disseminated with the aerosol, so it is of no concern in the context of contamination of environmental resources. The human population is currently highly exposed because the practice of vaccination has been interrupted.

Plague is caused by *Yersinia pestis* and can be transmitted through flea bites or in the aerosol. Also this agent is of no immediate concern.

TABLE 4.1. *A list of possible biological and chemical agents for terrorism. Each entry lists the normal route of exposure, when existing, and information on previous intentional uses.*

Definition	Agent	Possible origin	Normal route of exposure	Instances of terrorist threat	Dispersion route	Health effects
Biological hazards-algae (toxins)	Saxitoxin	Environment	Shellfish contamination		Food contamination	Paralytic or diarrhoeal poisoning
Biological hazards-bacteria	*Bacillus anthracis*	Occupational disease, cross-contamination from animals		2001 USA through post	Infectious aerosol with spores	Anthrax: inhalational, cutaneous, gastrointestinal
Biological hazards-bacteria	*Brucella:* B. melitensis (goats), B. suis (pigs), B. abortus (cattle), B. canis (dogs)	Diseased animals	Carry over to milk, only unpasteurised products; contact, inhalation		Aerial dispersion, skin abrasions	Brucellosis
Biological hazards-bacteria	*Burkholderia mallei*	Contaminated soil or water, infected animals		WWII	Aerial dispersion	Glanders
Biological hazards-bacteria	*Burkholderia pesudomallei*	Contaminated soil or water		WWII	Aerial dispersion	Melioidosis
Biological hazards-bacteria	*Coxiella burnetii*	Infected animals	Contaminated milk, faecal contamination, air dispersion		Aerial dispersion, food contamination	Q fever

(continued)

TABLE 4.1. (continued)

Definition	Agent	Possible origin	Normal route of exposure	Instances of terrorist threat	Dispersion route	Health effects
Biological hazards-bacteria	*Francisella tularensis*	Zoonosis	Insect bite, food-water con-tamination, inhalation	WWII	Aerial dispersion, food water contamination	Tularemia, rabbit fever
Biological hazards-bacteria	*Rickettsia prowazekii*		Arthropod vectors from infected humans		Food contamina-tion, infected vectors	Typhus fever
Biological hazards-bacteria	*Salmonella* sp., *Salmonella typhi*	Gastrointestinal tract of humans, live-stock species, dairy environ-ment, water, manure, workers	Faecal contamina-tion also during processing, only unpasteurised products	Salad bar1985 Oregon–Japan in the1960s	Dried bacteria sprinkled on food	Typhoid fever
Biological hazards-bacteria	*Shigella dysenteriae*, *Shigella* spp.	From workers, water	During cultivation of plants, cross-contamination, poor hygiene	1996, Laboratory workers	Food contamina-tion	Dysentery, shigel-losis
Biological hazards-bacteria	*Vibrio cholerae*	Water	During cultivation		Contaminated food water	Cholera
Biological hazards-bacteria	VTEC *Escherichia coli* O157	In animals digestive tract, water, manure, workers	Cultivation of plants, process-ing of foods, carry over to milk, only unpasteurised products		Food contamina-tion	Bloody diarrhea and abdominal cramps

Biological hazards-bacteria	*Yersinia enterocolitica*	In animals digestive tract	Carry over to milk, only unpasteurised products		Food contamination	Yersiniosis
Biological hazards-bacteria	*Yersinia pestis*	Rats	Flea vectors from rodents, direct contact	WWII	Droplets or aerosol, infection of rodents, contamination of objects, human contact	Plague (pneumonic)
Biological hazards-bacteria (toxins)	*Clostridium botulinum* neurotoxins		Canned foods, inadequate processing		Aerosol toxin, food water contamination with toxin or bacteria	Botulism
Biological hazards-bacteria (toxins)	*Staphylococcus aureus* (and others) enterotoxins	Contaminated food	Raw material or food handling, cross contamination		Aerosol toxin (SEB), food contamination	Diarrhoeal food poisoning
Biological hazards-fungi	*Coccidioides immitis*	Soil	Inhalation of conidia		Aerial dispersion	Coccidioidomycosis
Biological hazards-fungi (toxins)	Aflatoxins (*Aspergillus flavus*, *A. parasiticus*, others): AFB1, OTA, AFM1	Fungal contamination of feed and environment	Environmental contamination of food, metabolite in animal products; mycoflora on food products		Food contamination	Aflatoxicosis
Biological hazards-other organisms	*Ascaris suum*		Ova in food	Canada1970	Ova in food	Ascaris pneumonia, asthma, eosinophilia, pulmonary infiltrates

(continued)

TABLE 4.1. (continued)

Definition	Agent	Possible origin	Normal route of exposure	Instances of terrorist threat	Dispersion route	Health effects
Biological hazards-other organisms (toxins)	Ricin	Castor oil beans	Ingestion, inhalation	KGB assassination	Aerial dispersion, food contamination	Gastroenteritis
Biological hazards-protozoa	Giardia lamblia	Water, manure	During cultivation		Food contamination	Giardiasis
Biological hazards-virus	Alphaviruses–equine encephalitis viruses VEE EEE WEE	infected horses	Mosquito vectors from horses		Aerial dispersion, infected arthropods	Encephalitis
Biological hazards-virus	Bunyavirus (Rift Valley)	Zoonotic	Mosquito vectors		Aerial dispersion, infected arthropods	Fever, encephalitis
Biological hazards-virus	Filoviruses and arenaviruses (Ebola marburg Lassa Junin)	Zoonotic	Arthropod bites, rodent contamination of food, inhalation		Aerial dispersion, infected arthropods, contamination of objects	Viral hemorrhagic fevers
Biological hazards-virus	Flaviviruses – Tickborne and others	Infected animals	Mosquito vectors, ticks		Aerial dispersion, infected arthropods	Encephalitis, dengue, tick-borne complex encephalitis
Biological hazards-virus	HIV	Infected humans	Direct contact		Human contact, inoculation	AIDS
Biological hazards-virus	Nipah virus. hantavirus	Infected animals	Contact with infected pigs		Aerial dispersion, infected arthropods	Encephalitis

Biological hazards-virus	Variola major	Not possible	Only in laboratories	Historical, WWII	Aerosol, inhalation–human contact	Smallpox
Chemical hazards-inorganics	Arsenic hydride AsH	Mining, manufacturing	Occupational exposure		Inhalation	Blood agent, hemolysis
Chemical hazards-inorganics	Chlorine	Industries, manufacturing plants	Inhalation	WWI	Aerial dispersion	Asphyxiant gas
Chemical hazards-inorganics	Fluorine, hydrogen fluoride	Industries, manufacturing plants	Occupational exposure		Inhalation, ingestion, contact	Pulmonary irritant, eyes, mucoses, skin
Chemical hazards-inorganics	Heavy metals (arsenic, mercury, cadmium)	Environmental contamination of feed or food, water, manure	Plants uptake, surface contamination, accumulation in animal tissues	1978 Piking of oranges Hg - England 1984, spiking turkey Hg - Canada 2000, As in coffee	Spiking of food and water	Poisoning, carcinogenesis, teratogenesis
Chemical hazards-inorganics	Hydrogen chloride	Industries, manufacturing plants	Inhalation		Aerial dispersion	Blood gas
Chemical hazards-organics	Lewisite I				Aerial dispersion	Vesicant, blistering irritating eyes lungs

(continued)

TABLE 4.1. (continued)

Definition	Agent	Possible origin	Normal route of exposure	Instances of terrorist threat	Dispersion route	Health effects
Chemical hazards-organics	Phosgene	Industries, manufacturing plants	Inhalation	WWI	Aerial dispersion	Asphyxiant gas, eye irritant
Chemical hazards-organics	Sarin			Tokyo subway 1995	Aerial dispersion	Nerve gas
Chemical hazards-organics	Soman				Aerial dispersion	Nerve agent
Chemical hazards-organics	Sulfur/nitrogen mustard			WWI	Aerial dispersion	Vesicant, blistering eyes skin lungs
Chemical hazards-organics	Tabun				Aerial dispersion	Nerve agent
Chemical hazards-organics	VX				Aerial dispersion	Nerve gas
Chemical hazards-organics	Cyanogen chloride	Vegetables, industries		WWII	Aerial dispersion	Blood gas
Biological hazards-bacteria	Bacillus anthracis	Animal disease	Infection		Animal contamination	Anthrax of livestock
Biological hazards-virus	AHS virus – orbivirus	Animal disease	Infection		Animal contamination	African horse sickness of horses, mules, donkeys

Biological hazards-virus	AI virus–influenza virus	Animal disease	Infection		Animal contamination	Avian influenza of birds
Biological hazards-virus	Aphthovirus	Animal disease	Infection		Animal contact, contaminated objects	Foot-and-mouth disease of farm animals
Biological hazards-virus	ASF virus–iridovirus	Animal disease	Infection		Animal contamination	African swine fever of pigs
Biological hazards-virus	BT virus–reovirus	Animal disease	Infection		Animal contamination	Bluetongue of cattle, sheep, goats, wild ruminants
Biological hazards-virus	Calicivirus	Animal disease	Infection	New Zealand 1997	Animal contamination	Rabbit hemorrhagic disease
Biological hazards-virus	END virus–paramyxovirus	Animal disease	Infection		Animal contamination	Exotic Newcastle disease of birds
Biological hazards-virus	Enterovirus	Animal disease	Infection		Animal contamination	Swine vesicular disease
Biological hazards-virus	LSD virus–capripox-virus	Animal disease	Infection		Animal contamination	Lumpy skin disease of cattle
Biological hazards-virus	Morbillivirus	Animal disease	Infection		Animal contamination	Peste des petits ruminants, on sheep and goat

(continued)

TABLE 4.1. (continued)

Definition	Agent	Possible origin	Normal route of exposure	Instances of terrorist threat	Dispersion route	Health effects
Biological hazards-virus	Pestivirus	Animal disease	Infection		Animal contamination	Classical swine fever of pigs
Biological hazards-virus	RP virus–morbillivirus	Animal disease	Infection		Animal contamination	Rinderpest in wild animals
Biological hazards-virus	RVF virus–phlebovirus	Animal disease	Infection		Animal contamination	Rift Valley fever of livestock
Biological hazards-virus	SGP virus–capripox-virus	Animal disease	Infection		Animal contamination	Sheep pox and goat pox
Biological hazards-virus	VS virus–vesiculovirus	Animal disease	Infection		Animal contamination	Vesicular stomatitis pf equidae, bovidae, suidae
Biological hazards-fungi	*Bipolaris maydis*	Plant disease	Infection		Plant contamination	Southern corn leaf blight
Biological hazards-fungi	Phytophthora infestans	Plant disease	Infection		Plant contamination	Potato blight
Biological hazards-fungi	*Puccinia graminis* f. sp. tritici	Plant disease	Infection		Plant contamination	Wheat stem rust
Biological hazards-fungi	*Puccinia* spp.	Plant disease	Infection		Plant contamination	Cereals stem rust, on wheat, oat, barley
Biological hazards-fungi	*Pyricularia grisea*	Plant disease	Infection		Plant contamination	Rice blast
Biological hazards-fungi	*Tilletia indica*	Plant disease	Infection		Plant contamination	Wheat cover smu
Biological hazards-fungi	*Xanthomonas axonopodis*	Plant disease	Infection		Plant contamination	Citrus canker

Botulinum toxins, the most toxic compounds known, are typical food contaminants, produced by *Clostridium botulinum*. It is reported that in the 1970s a terrorist group planned the introduction of botulinum toxin in the Chicago water supply system (Sobel et al. 2002).

Staphylococcal enterotoxin B is also a food contaminant.

Francisella tularensis causes tularaemia and can be transmitted by ingestion of contaminated food and water, besides by inhalation and insect bites.

Other bacterial agents of relevance may be *Brucella, Burkholderia* (or *Pseudomonas*) *mallei, Coxiella burnetii*. Viral agents include Venezuelan equine encephalomyelitis, viral hemorrhagic fevers, Ebola, Marburg, Dengue fever, Hantavirus, Argentine hemorrhagic fever. Finally, other toxins include ricin, saxitoxin, shiga toxin, mycotoxins.

Pathogens could also be used to contaminate the production of food, instead of food itself: mad cow disease, blue tongue, brucellosis, food and mouth disease are examples of diseases which could be easily spread among food-producing animals (Table 4.1).

3.3. POSSIBLE SENSING TECHNIQUES

Monitoring techniques and sensors are currently available for deployment at critical sites. This paper does not exhaustively complete a list, but only some examples.

To monitor physical hazards, sensors can measure temperature, moisture, particle size, radioactivity, positioning and orientation (GPS), remote sensing with aerial devices.

For monitoring chemical hazards: inorganic compounds can be assessed with voltammetry, spectroscopy, spectrometry; volatile compounds are measured with gas chromatography, gas sensors, electronic noses; organic compounds can be assessed with liquid or gas chromatography, spectroscopy and spectrometry. Sensors based on biological recognition can detect chemical hazards. Enzyme based assay, like ELISA, is the most classic of immunological techniques, advantageous because it amplifies a signal. A large series of assays are available for several organic pollutants; their sensitivity and efficacy is comparable with those of classical analytical techniques: liquid chromatography- or gas chromatography-mass spectrometry. ELISA tests can be formulated also as dipstick or lateral flow assays. Environmental contaminants are usually present in complex matrices and mixtures, so their detection requires a phase of extraction and purification followed by sensitive analytical methods for quantification. Innovation therefore requires the development of sensitive and robust techniques for evaluation of several pollutants at once, possibly in automated way without previous treatment.

Monitoring of pathogenic microorganisms is mainly performed off-site. The organism captured from the environmental substrate of interest is brought to the laboratory and there analysed (Arora et al. 2006) by means of:

- Microscope observations
- Culturing techniques on selective media
- Genetic methods based on molecular markers
- Immunological techniques
- Gas chromatography or HPLC for specific compounds

All these methods require a long sample preparation, and are time-consuming for the analysis (days). Moreover, culture-based methods disregard all the forms of viable but non culturable microorganisms.

Biological hazards can alternatively be detected with methodologies based on immunology, enzymatic assays, proteomics, metabolomics and genomics.

Several innovative devices have been recently developed for monitoring terrorism in high-risk locations, using the analysis of nucleic acids or proteins (Regan et al. 2008). The Autonomous Pathogen Detection System (APDS) concentrates periodically airborne particles (1–10 μm) into a liquid; then it performs a multiloci detection of several biothreat agents with PCR followed by hybridisation to pathogen-specific microspheres and recognition with fluorescence. The system operates autonomously for a week performing seven analyses per day.

New monitoring capabilities are continuously being developed, especially by increasing the multiplexing possibilities. DNA microarrays give the possibility of challenging thousands of DNA sequences at the same time against a sample, and the pattern of hybridisation will provide information about the organisms present in the sample. A microarray system has been developed for detection of several pathogens of potential interest in bioterrorism: *Listeria, Campylobacter, Staphylococcus aureus, Clostridium perfringens* (Sergeev et al. 2004).

Miniaturisation is bringing to the field new and complex tools such as surface plasmon resonance, Real Time PCR and chromatography. Real Time PCR machines can perform DNA amplification and monitor the course of the reaction by measuring fluorescence. Portable PCR machines have been developed which can be deployed in the field, near the critical points. Application for biodefense is amongst the most stimulating. A test for seven different biological agents has been developed and tested on different portable PCR machines which performed with comparable sensitivity (Christensen et al. 2006).

The lab-on-chip is an application of microfluidics, a technology dealing with capillaries and small amount of fluids, ranging from 10^{-9} to 10^{-18} L. Lab-on-chips may contain chambers for PCR amplification, DNA hybridisation, electrophoresis. One important reason behind the development of microfluidics systems in the 1990s was the search on sensors which could be employed in the field to detect chemical and biological agents (Whitesides 2006).

4. The SITCEN project

The NATO Science for Peace and Security program (www.nato.int/sps/index. html) aims to stimulate interactions among Institutions from partner countries and from NATO countries on themes concerning scientific endeavours such as combating pollution and counteracting threats to the environments, both accidental and intentional. In this framework, a project was launched in 2007 from a partnership between the University of Parma-Consortium for Environmental Sciences (Italy), and the Academy of Geopolitical Problems (Russian Federation): "Development of a prototype for the International Situational Centre on Interaction in Case of Ecoterrorism" (SITCEN).

The background of the project considers the increase witnessed in past years in the number and in the magnitude of intentional threats towards the environment. We must assume that environments in rural settings and in anthropogenic settings will be more and more threatened by terrorists. As described, many facilities and infrastructures are vulnerable to attacks: water supply, agricultural soils, agricultural deposits of chemicals, crops and farms, urban infrastructures, factories. The strategic importance of menacing vital infrastructures and resources has been recognised and widely applied as in warfare. In addition, recent instances of accidental contamination of vital resources have showed us the impact that this exerts on citizens and consumers. If we postulate that an event like the Chernobyl explosion or the oil spill from the Exxon Valdez were caused intentionally, rather than accidentally, it is not far from true to reckon how damaging this could be for governments and citizens. More, epidemics like the BSE (mad cow disease) or the avian influenza could be provoked by terrorists with appropriate biological agents. These examples have shown to the public opinion how governments were not ready with countermeasures against accidents of this kind. There is a lack of coordination and interaction at international level, whereas this should be essential in cases of terrorism against environmental resources, because experience has shown how damage to the environment is rarely circumscribed to a single country.

From these examples it becomes evident that there are gaps in the surveillance capability simply because of the number and extension of potential targets. Surveillance, control, monitoring, are at the moment concentrated on critical areas of potential highest impact, and currently there are not enough resources for monitoring other types of environment. However, it can be recognised that many of the potential threatening agents, of chemical and biological origin, could be detected and identified by means of existing surveillance networks which monitor environmental resources and agricultural assets. What is missing is rather a coordination of these efforts, such as to collect and organise data coming from different sources in a comprehensive way.

The SITCEN project was aimed to develop a prototype of a Situational Center which could provide this missing role. The Situational Center could collect information from several different sources, providing a framework and infrastructure in which data could be organised, analysed in a meaningful way, and used to provide support to decision-makers. Such an ICT infrastructure should also favour training and capacity building to users from different countries, provided that a link is placed between the Situational Centers and any National Center involved in collecting data at different regional levels.

Databases for organising data and connecting them to geographical information systems (GIS) are very important in this type of Center. Only after rationalisation and classification the data can be used to recognise anomalies, give early warnings and assist in taking decisions. Authorities of countries involved must be able to access the information in databases and must also access facilities for data analysis. In case of transboundary problems, cooperation among countries will require access to the same information sets.

The project has been developed to address three kinds of substrates, or environmental resources: (i) water, inland water, freshwater reservoirs, municipal facilities and utilities, recreational sites; (ii) air, especially in highly anthropogenic environments; (iii) agricultural environments, especially soil, impacting on plant cultivation and animal farming.

The types of risk and hazardous agents which can affect these substrates can be: (i) chemical, (ii) biological, and (iii) physical. Chemical agents include toxic contaminants like asphyxiant gases, vesicants, nerve agents, blood toxicants, usually disseminated through air and aerosol or dissolved in water. Biological agents include organisms pathogenic for humans, animals and plants, from all biological groups: viruses, bacteria, fungi, protozoa. Also toxins produced by some of these organisms can be used as such in contamination. Biological agents can be used to contaminate food products, or to attack crops and farm animals, or even be disseminated in air and water. Finally, physical agents include fire, radioactive materials, high pressure, and others which can compromise integrity of the environments.

The SITCEN project considers these three environments and relative risks in the context of intentional threats and attacks in what we term "ecoterrorism". It is bringing to the development of an International Situational Center which (i) receives and stores information, (ii) develops models and simulations, (iii) provides easily interpretable representation of data, (iv) prepares guidance documents. The project lasts for 3 years, from April 2007 to April 2010.

The project participants have worked on two main aspects: (i) mapping and characterisation of risks, and (ii) building up the Situational Center.

For the first task, the partners have collected exhaustive information from literature and publications, concerning all possible chemical, physical and biological agents of potential application in terrorism threats against environmental resources. Some examples of vulnerability points in typical food supply chains have been provided: just as an instance, in the tomato chain vulnerabilities may concern the growth of plants in the nursery and greenhouse, where they can be subjected to chemical contamination by heavy metals, radionuclides, organic contaminants, or they may concern the harvesting, collecting and distributing phases, open to contamination with bacterial or viral pathogens. The partners have also collected available information on analytical techniques for detection of biological hazards, bacteria and other microorganisms, based on molecular biology and application of biosensors. Research carried out at the University of Parma provides interesting examples of bacteria isolation and detection with Polymerase Chain Reaction and DNA microarrays (Marmiroli and Maestri 2007; Marmiroli et al. 2008).

The task of building the International Situational Center was carried out in Moscow, in cooperation with the Russian Academy for Public Administration and with the company ARTI. The information flows from the different National Centers (for information gathering) to the International Situational Center (ISC) where the data will be processed, stored and analysed. From the ISC the information processed (or integrated) can be retrieved, for use at the national (regional) level. Clear guidelines are required about collecting, storing and disseminating information, safeguarding confidentiality and sensitivity of data. Data from environmental sources are georeferenced, stored in the databases of a Geographical Information System. It allows representation of data on maps, leading to a comparative visualisation. The realisation of a proper GIS will allow a meaningful analysis of data in connection with their geographic relations, necessary when environmental resources have to be described.

Another important feature of the SITCEN project consists in the efforts towards capacity building and training, aimed towards increasing awareness around ecoterrorism by disseminating information at many levels: from students enrolled in graduate and post-graduate course, to end-users in companies and public administrations, to scientists in conferences and meetings.

5. Conclusions

In listing countermeasures against terrorism, information and awareness are surely among the most important. Ignorance about risks and effects is an additional weapon in the hands of terrorists. The experience of the last years has shown that initiatives like those financed by NATO Public Diplomacy Division have encouraged cooperation and interaction between the stakeholders interested in all aspects of science and technology applied as countermeasures against terrorism, helping in spreading knowledge and information. Projects, conferences, meetings, summer schools have involved hundreds of participants and brought to production of texts and books. Particularly significant, in 2006, the COST – NATO Strategic Expert Meeting on Food Security and Simulation (www.cost.esf.org) brought together 45 participants from 15 countries to discuss gaps in research and areas for cooperation. Coordination among bodies working in this direction, including the EC Framework Programmes for research, will become more and more essential to increase preparedness.

In this year 2009 marking its 60th anniversary, the role of NATO in promoting Peace through Science can be recognised as highly valuable for achieving international consensus.

Acknowledgements

The authors acknowledge funding from NATO's Public Diplomacy Division for Short-Term ad hoc project ECOTER and in the framework of "Science for Peace", project SFP982498 SITCEN. Contributions have also been provided by GSA project MENTORE and by the University of Parma.

Bibliography

Anderson, F., 2003, Security and Water. In: Water:Science and Issues. The Gale Group Inc. Retrieved February 23, 2009 from Encyclopedia.com: http://www.encyclopedia.com/doc/1G2-3409400301.html

Arora K, Chand S, Malhotra BD (2006) Recent developments in bio-molecular electronics techniques for food pathogens. Anal Chim Acta 568:259–274

Bean, N.H., Griffin, P.M., Goulding, J.S., and Ivey, C.B., 1990, Foodborne disease outbreaks, 5-year summary, 1983–1987, *MMWR Morb. Mortal. Wkly. Rep.* 39(SS01):15–23

Carus, W.S., 1998, Bioterrorism and biocrimes. The illicit use of biological agents since 1900, Center for Counterproliferation Research, Washington D.C

Chalk, P., 2004, Hitting America's soft underbelly. The potential threat of deliberate biological attacks against the U.S. agricultural and food industry, RAND Corporation, Santa Monica CA [www.rand.org]

Chalecki EL (2002) A new vigilance: identifying and reducing the risks of environmental terrorism. Global Environ Politics 2:46–64

Christensen DR, Hartman LJ, Loveless BM, Frye MS, Shipley MA, Bridge DL, Richards MJ, Kaplan RS, Garrison J, Baldwin CD, Kulesh DA, Norwood DA (2006) Detection of biological threat agents by Rea-time PCR: Comparison of assay performance on the R.A.P.I.D., the LightCycler, and the Smart Cycler platforms. Clin Chem 52:141–145

European Commission, 2006, Avian Influenza. Health & Consumer Protection Directorate-General. Retrieved February 23, 2009 from ec.europa.eu/food/animal/diseases/controlmeasures/avian/ai_factsheet_2006_en.pdf

European Commission, 2007, Commission Regulation /EC) No 1441/2007 of 5 December 2007 amending Regulation (EC) No 2073/2005 on microbiological criteria for foodstuffs. OJ L. 322, 7.12.2007, pp. 12–29

Guillemin, J., 2006, Scientists and the history of biological weapons, *EMBO Rep.* 7, Special Issue S45–S49

Mac Kenzie WR, Hoxie NJ, Proctor ME, Gradus MS, Blair KA, Peterson DE, Kazmierczak JJ, Addiss DG, Fox KR, Rose JB, Davis JP (1994) A massive outbreak in Milwaukee of Cryptosporidium infection transmitted through the public water supply. N Engl J Med 331:161–167

Manning L, Baines RN, Chadd SA (2005) Deliberate contamination of the food supply chain. Br Food J 107:225–245

Marmiroli, N., and Maestri, E., 2007, Polymerase Chain Reaction (PCR), in: *Food Toxicants Analysis. Techniques, Strategies and Development,* Y. Picò, ed., Elsevier, Amsterdam (The Netherlands), pp. 147–187

Marmiroli, N., Palumbo, G., Consigli, C., Agrimonti, C., Sanangelantoni, A., and Maestri, E., 2008, Innovative tools for detection and enumeration of contaminant micro-organisms in the poultry food supply chain, Proceedings of 42nd International Symposium on "Analytical Technologies: Tools and Implementation Strategies in Animal Science", Porto Conte Ricerche, Alghero, Italy, May 2007 pp. 131–142

Pellerin C (2000) The next target of bioterrorism: your food. Environ Health Perspect 108: A126–A129

Regan JF, Makarewicz AJ, Hindson BJ, Metz TR, Gutierrez DM, Corzett TH, Hadley DR, Mahnke RC, Henderer BD, Breneman JW IV, Weisgraber TH, Dzenitis JM (2008) Environmental monitoring for biological threat agents using the autonomous pathogen detection system with multiplexed polymerase chain reaction. Anal Chem 80:7422–7429

Sergeev N, Distler M, Courtney S, Al-Khaldi SF, Volokhov D, Chizhikov V, Rasooly A (2004) Multipathogen oligonucleotide microarray for environmental and biodefense applications. Biosens Bioelectron 20:684–698

Sobel J, Khan AS, Swerdlow DL (2002) Threat of a biological terrorist attack on the US food supply: the CDC perspective. Lancet 359:874–880

Stamm BH (2002) Terrorism risks in rural and frontier America. IEEE Eng Med Biol Mag 21:100–111

Van Larebeke N, Hens L, Schepens P, Covaci A, Baeyens J, Everaert K, Bernheim JL, Vlietinck R, De Poorter G (2001) The Belgian PCB and dioxin incident of January-June 1999: exposure data and potential impact on health. Environ Health Perspect 109:265–273

Whitesides GM (2006) The origins and the future of microfluidics. Nature 442:368–373

World Health Organization (WHO) (2004) Public health response to biological and chemical weapons – WHO guidance. World Health Organization, Geneva

World Health Organization (WHO) (2008) Terrorist threats to food: guidance for establishing and strengthening prevention and response system. World Health Organization, Geneva

CONSUMER TRUST IN FOOD UNDER VARYING SOCIAL AND INSTITUTIONAL CONDITIONS

UNNI KJÆRNES

*The National Institute for Consumer Research (SIFO),
Oslo, Norway
e-mail: Unni.kjarnes@sifo.no*

Abstract: Consumer trust in food has been on the agenda in Europe over the last 10-15 years, brought forth by major events linked to food safety and quality as well as by structural and political changes in the food system. Research has shown that consumer responses cannot be reduced to a matter of unbalanced media presentations ('scares') or personal uncertainty in view of technological innovations (like GM food). Such aspects are certainly important, but the key factor triggering responses of distrust seems to be how market and public actors handle such issues. Lack of accountability and transparency and disregard of consumer expectations and interests form important explanations to negative reactions among buyers and eaters of food. Media presentations are important in communicating such problems, which of course also may be exaggerated. But lack of openness and responsiveness seems to be more of a problem than media amplification.

Public food control was early installed in order to handle problems of safety and accountability. From the late 1990s, food provisioning systems changed dramatically and public control was modified in order to cope with these changes. More market integration, concentration and complexity, and regulation based on market self governance did not only question consumers' trust; trust also became more important. Gradually, self-governance systems have improved and public regulatory efforts have become clearer and more sensitive to consumer expectations. In that way, new divisions of responsibility have been established which also give people in their capacity as consumers a (somewhat) more visible role.

Or-at least - this seems to have happened for food safety. Several other food issues are less settled. And it has happened in some European countries, while other countries still struggle with considerable scepticism both among ordinary consumers and in other actors' attitudes towards consumers of food. An important lesson is that trust in regulatory regimes largely based on market self governance depends on their ability to demonstate predictability and accountability and their attentiveness towards public and consumer concerns and socially agreed standards. The whole system must meet basic expectations of fairness and legitimate distribution of responsibility. When such conditions are met, the capability to build trust seems considerable.

Yet, all of this has taken place within a context of relative affluence, abundant supplies with decreasing food prices, and rather stable institutional contexts. A number of recent events suggest that these conditions may not be taken for granted in the future. Market instability, increasing food prices, more opportunistic behaviour among market actors, and less concern and legitimacy for consumer expectations, are all possible scenarios. This may, in turn, challenge established regimes of food regulation and (again) make food issues more contentious. Historically, it is in periods of rapid change that we find most cases of consumer mobilisation and food riots. But responses among consumers and the public do not necessarily depend on the emergence of new conditons or problems *per se*, but on whether these challenges are met in ways that consider consumer concerns and interests.

Keywords: Food safety, consumer trust, food security

1. Changes in the food sector and popular trust

The expansion of food markets from the sixteenth century onwards and the emergence of modern food industries from the nineteenth century have been accompanied by public regulatory interventions (Lyon 1998; Green et al. 2003). Measures to protect the population and concerns for producers and commerce have been negotiated in different ways and by a variety of means. Public food control was early installed in order to handle problems of safety and accountability and food prices have been of concern both with regard to poverty and preparedness for times of crisis and short supplies. While stable supplies and technological innovation have helped to produce better predictability, people are well aware that business concerns for consumer needs and expectations will inevitably be balanced against profitability. Moreover, competitive commercial actors will have a tendency not to be too open about their own actions. Regulatory intervention has been crucial for establishing accountability as well as meeting consumer demands that are not easily commercialised (market externalities). The inherently contingent character of trust in food is thus linked to the willingness of market actors to adhere to societal norms and consumer expectations and to keep their promises.

In the long run, the dynamic interrelationship between market expansion, new food technologies and regulatory institutions have brought more stable supplies, more variety and more predictable quality. And, generally speaking, stability and predictability seem to be nurturing a basic confidence in the food that is marketed. Popular trust in food may be stable or not even thematized for extended periods of time. Many people, as buyers and eaters of food, have come to believe that producers and retailers will meet their expectations and take sufficient consideration of their interests and needs referring to a range of issues; first of all safety, quality and price, at times even issues not directly associated with the food products as such, for example labour conditions.

Since the late 1990s, food provisioning systems have changed dramatically and public control has been modified in order to cope with these changes. Not only have food markets expanded and new, advanced technologies have been introduced. Power and control have also been subject to dramatic transformations; with growing complexity developing along with more market integration and concentration, especially at the retail level. At the same time, people's expectations seem to have changed. The established authoritative expert based bodies, serving consumers' interests in a rather closed and paternalist fashion, were increasingly challenged. The GM controversy and others, like BSE, salmonella, etc., demonstrated the inadequacy of established systems, not the least because also more and more consumers wanted to have a say. Along with market changes and the politicisation of consumer interests, we have seen regulatory reforms which have emphasised science based risk regulation, international harmonisation, market self governance, and public (or other third party) audits (Halkier and Holm 2006). These reforms have been instigated based on pressure from large business corporations or producer groups combined with concerns over consumer, as indicated by the emphasis in transparency and independent risk assessment.

In that way food supply is more actively controlled than ever. But the changes have at the same time meant more imbalances of power and information between suppliers and consumers. In a transition phase, these changes did not only challenge consumers' trust; trust also became more important. One major answer has been more focus on labelling, allowing people to choose and thus make their voice heard – which is important. But consumers have no direct command over food and how it is being produced. Mishaps do occur and with more complex systems the impacts of these mishaps may increase significantly. People are also highly aware that there are black sheep, sellers who try to earn quick money by mislabelling or otherwise behaving unaccountably. Gradually, self-governance systems have improved and public regulatory efforts have become clearer and more sensitive to consumer expectations. In that way, new divisions of responsibility have been established which also have given people, in their capacity as consumers, a (somewhat) more visible role.

Or - at least - this seems to have happened for food safety. Several other food issues are less settled. And the process of settling conflicts and uncertainties has been more successful in some European countries, while other countries still struggle with considerable scepticism. An important lesson is that trust in regulatory regimes largely based on 'governance at a distance' depends on the ability to demonstrate predictability and accountability and the attentiveness towards public and consumer concerns and socially agreed standards. The whole system must meet basic expectations of fairness and legitimate distribution of responsibility. When such conditions are met, the capability to build trust seems considerable.

But responses are not only about predictability and accountability in food provisioning systems. Standards are becoming increasingly important and a lot of controversy has addressed these standards are set and who is setting them. Again we must consider the huge imbalances of power, information and knowledge between providers and consumers of food. There is no way that individual consumers can compete with that. So there need to be organisational solutions are needed that ensure consideration of consumers' interests.

2. Variations in trust depend on institutional and cultural conditions

Mechanisms of trust are therefore linked to culturally founded expectations as well as institutional set-up and performance (Mishler and Rose 2001; Seligman 1997; Tilly 2004). Considering these contingencies it is therefore not surprising that we find large variations in the levels and sources of trust in food (as well as trust referring to other issues) across Europe. Even though good data are lacking, it is also likely that changes have taken place over time, sometimes in the form of rapid shifts, reflecting how problems and conflicts have been handled by public authorities and market actors.

A public opinion survey conducted in six West European countries in 2002 in the project TRUSTINFOOD found that the Scandinavians and the British are generally trusting, while much more distrust in food is found in countries like Germany, Portugal and Italy (Kjærnes et al. 2007). This is for example reflected in public opinions on the safety of various food items. Whereas Europeans tend to agree that less safety problems are associated with vegetables than meats and people tend to worry less about fresh than processed foods, the overall levels of confidence in food safety in various countries are highly diverging. Moreover, people in the low-trust countries are considerably more pessimistic about changes in the food sector over the last 20 years, compared to the high-trust regions of North-West Europe. It is noticeable that this pessimism extends to a range of food issues, including food prices, quality, nutrition, safety and methods in food production. In fact, Europeans (in 2002) tended to be more worried about changes in quality and price than they were about food safety.

As a third indicator of trust, we asked how people viewed the truth-telling of various institutional actors, whether they were expected to tell the whole truth in case of a scandal with salmonella in chicken, parts of the truth or rather withhold information. The answers are striking in two respects. The ranking of institutional actors was identical across Europe; civil society actors like experts and consumer organisations were most often expected to tell the whole truth, public authorities were in the medium range together with the media, while all market actors, including farmers, processors and retailers, were generally expected to withhold information. This was the case even for politicians. The second point is that the country differences are reproduced in the aggregate effects regarding the truth-telling of institutional actors; Italians, Portuguese and German respondents tend to be generally more sceptical towards public as well as private actors in the food sector than the Norwegians and the Danes, in particular, but even compared to the British. Many of these features appeared in a Eurobaromenter survey on food safety from 1998 (European Commission 1998). A public opinion survey on animal welfare issues in relation to food consumption in 2005 also shows more or less the same patterns of country variations in trust in food (Kjærnes and Lavik 2007).

Individual psychological mechanisms and family patterns of eating are certainly important for how we relate to and deal with food in our everyday lives. Individual anxieties and preferences, learning of competencies, and concerns for our children's safety, all matter (Hansen et al. 2003). Such features are influenced by life cycle, status and resources. We might therefore expect considerable variation depending on factors like family situation, education level, and age. However, looking across Europe, we are unable to identify major systematic variations in trust between socio-demographic groups. Or, more precisely, time and country of residence matter more. There are variations depending on social stratification, age and household type, but these are not aligned across various countries (Kjærnes et al 2007). The only consistent finding is that women worry more than men. This is a well-known feature of all questions related to food purchases, preparation and eating. Women take on more responsibility and they worry more. This general finding is not very helpful in explaining why trust varies and changes but it is of course a reminder when it comes to measures addressing food consumers.

There are no simple reasons for the variable national patterns of trust. There are some deep-set cultural factors about trust in other people. But what seems to be the case is that high-trust countries generally have configurations of powerful actors, including various market actors and food authorities, who *together* have managed to produce some clarity and consensus over responsibility and control of food. In low-trust countries we find more controversy about food issues, less coordination and clarity, fragmentation of responsibilities, and lack of transparency. Britain experienced a crisis of trust in food in the late 1990s, but new systems were put in place which seem to have allayed the distrust – at least for a time. One background may be that even though many British consumers do question concrete actors in the food system, they have a basic belief in the workability of impersonal, institutionalised "systems" such as food standards and control procedures.

This is however not only about institutional performance. Political culture and the role of consumers are also important. Instead of seeing food scandals as a source of consumer distrust, problems in the food sector seem to turn into scandals more easily in some places (and during some periods) than in others. Again using Great Britain as an example, this country has a political culture emphasising controversy and heterogeneity. Consumer involvement and responsibility are important. Norway, on the other hand, has a considerably more consensual political culture. Here consumers are primarily to be protected and do not represent any influential public or political voice. The German situation is generally more conflictual, where all actors in the food sector, including consumers, seem to be highly sceptical towards each other. Rather than seeking overall institutional solutions, therefore, many worried consumers rely more on personal ways of handling their distrust, for example by purchasing their food in small, special shops and via alternative provisioning channels. Emphasis on personal relations and local systems of provision is even more pronounced in the south of Europe. There consumers' trust in food is about familiarity and food quality (encompassing a range of aspects from taste to nutrition, safety and sustainability), more than referring to impersonal systems for ensuring food safety – which are less trusted.

3. Shifting times?

Recently, food security has re-entered the public agenda, reviving an aspect of consumer politics that historically has appeared in waves. The uncertainty of supplies and its association with social order was for centuries managed by public granaries. With nineteenth century liberalism, the granaries were generally dismantled. After shortages – and riots – following from trade blockades and reduced production during World War I, the ideas of public regulation in the form of food security policies regained popularity (e.g. rationing and price regulation). Regulatory systems were established during the inter-war period, later to be reinforced by needs during and after World War II. But, food security policies have since the 1980s and 1990s been changed away from strong public interventions. Without the same media or political attention, these policies seem to have gone further than food safety policies in turning from public regulation to reliance on the responsibility of private actors. A major motivation appears to be growing supplies and strongly improved distribution systems, logistically as well as technologically. The task of public policy-makers has thus become more distanced, producing early-warning systems, plans of how such private organisations may work in instances of emergency, information arrangements, etc. This is the case in most (or perhaps all) countries in the western parts of the world (expressed for example within the context of NATO preparedness programmes). Shortages are to be dealt with mainly by reallocation of supplies and the mobilisation of private resources. The effects on distribution between consumer groups with different purchasing power and the possibility of social unrest are thus becoming less central in these policies.

One reason why this regulatory shift has received less attention may be that all of this has taken place within a context of relative affluence, abundant supplies with decreasing food prices, and stable institutional contexts. A number of recent events suggest that these underlying conditions may not be taken for granted in the future, one important factor being environmental problems associated with climate change. Instability of supplies, increasing food prices, more opportunistic behaviour among market actors may challenge established regimes of food regulation and again make food issues more contentious.

Production failures have negatively affected small subsistence farmers, but the main victims of surging food prices are consumers with low purchasing power and no alternative sources of food. As in former periods with supplies lower than demand, this may represent a major impetus for higher food production. In the long run, as advocated in liberalist economic theory, trade has generally produced overall higher total outputs. This is relevant even in the current situation. But in real life, considering not only environmental consequences but even social challenges, outcomes are much more complex. Environmental issues are currently being discussed extensively. I will here mention some of the social challenges associated with food consumption and the role of consumers.

Historically, it is in periods of rapid change that we find most cases of consumer mobilisation and food riots. Still, responses among consumers and the public do not necessarily depend on the emergence of new conditions or problems per se, but on whether these challenges are met in ways that consider consumer concerns and interests. Beliefs in the fairness of chosen solutions are crucial. The recent experiences thus raise classical problems associated with insufficient supplies in open markets. As noted already by Adam Smith, the beneficial effects of markets are in the long run, while people's experiences and responses are always acute. Open markets will allocate supplies to consumer groups and areas with the highest purchasing power, not to those in most need. Historically, perceptions of injustice and speculation have repeatedly incited food riots (Coles 1978; Thompson 1971; Tilly 1975). This is also what we saw around the world during the winter and spring of 2008. And this is the classical as well as current background for a number of measures introduced to maintain social order and/or to satisfy basic needs by intervening in market based distribution. Based on extensive empirical studies, Dréze and Sen (1989) have claimed that crises related to satisfying needs of food are mainly about social distribution systems and, notably, about entitlements, rather than shortages per se. It is therefore of utmost importance that policies to ensure food security consider long-term as well as short-term effects.

The events over the last couple of years have demonstrated that while climate problems are global and interdependencies are world-wide, the immediate solutions are generally sought nationally. Path-dependencies with regard to relationships between citizens, the state, and food production and distribution will probably matter for how these challenges are met. Several countries, especially where large population groups are vulnerable to souring food prices, have introduced traditional measures, like price regulation and restrictions to trade. This, in turn, has raised concerns over the long-term effects of these responses.

Climate change seems to introduce a new kind of uncertainty where visions of the relationship between the promotion of trade and concerns for food security may need to be reconsidered. While markets and technologies are basically different, the lessons of public unrest and distrust are still valid, namely that the production side as well as the consumer side must be taken into account – and that these concerns are not necessarily aligned.

4. Consumer adaptations to change

The European food sector is characterised by its dynamism and interdependencies within Europe and as part of global food markets. As mentioned above, new uncertainties can be added. Considerable flexibility and adaptability may therefore be needed on the consumer side. The adaptability of food consumers is associated with a complex dynamics between trust and people's willingness, on the one hand, and their capability, on the other. And, as already mentioned, consumers seem more "jumpy" in some places than in others, associated with political culture, the behaviour of market actors, as well as consumer roles.

The adaptability depends on the character of household practices in terms of cultural openness to change; practical and theoretical competencies; and the potential mobilisation of resources related to incomes and alternative sources of food. We also have to consider the concrete everyday patterns of buying, cooking and eating food. These are habits that are generally taken for granted and not easily changed, but it seems likely that more open and varied patterns as well as culinary and cooking skills may enhance flexibility. Other factors include storage space and cooking facilities. From earlier episodes of social unrest as well as recent studies of food we know that some food items are more important than others – not only practically and nutritionally, but even symbolically. Shortages or safety problems associated with basic foodstuffs such as meat or bread/grain have provoked the strongest reactions. Household resources, habits and food cultures thus have a strong influence on the capability of consumers to adapt and to face problematic situations.

When it comes to responses of trust and distrust, they are not a product of problems with health hazards or food shortage per se, but are associated first of all with how responsible food institutions (public and private) handle the problems. Accountability and justice are key factors. More significant than in a calmer situation, people must be convinced that public and consumer interests are not offset by concerns for profits and prestige. People's experiences and expectations will have impacts. Many countries experienced during World War II that information campaigns about challenges to the food supply and established trust in involved institutional actors produced considerable consumer loyalty and willingness to adapt (Cohen 2000; Trentmann and Just 2006). Without these preconditions, problems may easily trigger considerable social unrest, including destructive hoarding of food.

A third element here is the social role and responsibility assigned to ordinary people in their capacity as food consumers. This will influence their capability as well as their willingness to adapt. Over the last couple of decades we have seen that while people's direct control over the food they acquire is drastically diminished, their responsibilities seem to have increased. One indication of that is the growing emphasis on conscientious choices. Product labelling and other information systems are put in place to enhance this process. Also, consumers have become considerably more visible in politics, in the mass media, and in social mobilisation on a range of social causes (e.g. animal welfare, child labour, environmental sustainability). This process is in many ways positive, opening up for people to have a say via their role as consumer. But a precondition is that people are given choices that are real and reliable and that their choices or other manifestations of consumer voice matter. Across Europe, we find considerable variation; from highly optimistic consumers in the UK, denouncement of responsibility in for example Norway and Germany, to strong feelings of powerlessness in some East European countries (Berg et al. 2005). Responsibility without power and influence, a situation that is far from unlikely, may generate protest or, worse, apathy.

Consumer protest may signify many different things; general distrust in the system, protest against the burdens and responsibilities that consumers are expected to accept, as well as – and this is important – positive participation and responsibility taking for solving social problems. In any case, it is highly advisable to take such responses seriously and to analyse what these protests or general dissatisfaction are about. If done properly, that kind of exercise may by itself enhance the legitimacy of institutional actors. In turn, this is an important precondition for people's flexibility and loyalty in a situation of rapid change.

5. Conclusions

Consumer trust are influenced by trade policies, regulatory arrangements, shifts in power in the food chain, technological development, and other large-scale tendencies of structural change. Consumer trust and distrust vary considerably and consistently and cannot be accounted for only by individual, psychological explanations. Mechanisms behind trust and distrust include combinations of general, long-term cultural conditions, on the one hand, and the institutional setup and performance of a given provisioning and regulatory system, on the other hand. Trust seems less dependent on consumers' own strategies or resources. Changes in trust are influenced by complex interrelations between these actors and arenas and cannot be limited to single media scandals or carefully engineered public measures (even though such factors may certainly have an impact). Findings from several studies suggest basic differences between the Nordic, British, South and Eastern European situations. Some countries, like Germany, seem to be in a transitional or positively conflictual state. European market and regulatory integration and processes of globalisation certainly have impacts, but these processes are being handled very differently across Europe, due to national economic interests, political traditions, etc. As a general lesson, trust in modern, advanced and rapidly changing food provisioning systems seems to require specific arrangements to recognise and handle consumer scepticism and distrust.

Trust phenomena are not static and consumer trust is not won once and for all. Issues causing concern may be shifting. What we are seeing today may be a shift in focus from safety to nutrition, quality and – perhaps – food security. Conflicts referring to a variety of issues may turn into political crises and again erode trust. As a more long-term tendency, we may see the emergence of more demanding and critical consumers who are active in the market and in public debates. In such a situation, consumers may turn out to be harder to satisfy, but also taking on more responsibility for various food issues.

What then if there is a situation where flexibility (especially with regard to the selection of food products) is being reduced and uncertainty increases? One quite likely kind of response is consolidation and loyalty, where people understand and accept that they as consumers have to contribute their share. However, this may be challenging. Instable supplies may mean that ordinary habits are difficult to maintain, a condition that by itself may produce uncertainty. As some people will be affected more than others, reactions will depend on how the situation is handled. There may be conflicts over the distribution of food as well as over safety risks. Institutional trust thus becomes more important but also more vulnerable. There is a lot to learn from history here, including distant as well as more recent experiences. But we know little about how these challenges will turn out under new regulatory regimes, with better systems for handling uncertainties if distribution but also more responsibility assigned to market actors and consumers. In the present situation, I think these issues should receive more attention.

Bibliography

Berg, L., U. Kjærnes, E. Ganskau, V. Minina, L. Voltchkova, B. Halkier, and L. Holm. 2005. "Trust in food safety in Russia, Denmark and Norway." European Societies 7 (1):103–30.

Cohen, L. 2000. Citizens and consumers in the United States in the century of mass consumption. In: M. Daunton and M. Hilton (eds.) The Politics of Consumption: Material Culture and Citizenship in Europe and America. Oxford: Berg.

Coles, A. J. 1978. "The moral economy of the crowd: Some twentieth-century food riots." Journal of British Studies 18 (1):157–76.

Dréze, J., and A. Sen. 1989. Hunger and public action. Oxford: Oxford University Press.

European Commission. 1998. Eurobarometer 49 on food safety. Brussels: European Commission DG XXIV and Inra European Coordination Office.

Green, K., M. Harvey, and A. McMeekin. 2003. "Transformations of food consumption and production processes." Journal of Environmental Policy and Planning 5 (2):145–63.

Halkier, B., and L. Holm. 2006. "Shifting responsibilities for food safety in Europe. An introduction." Appetite 47 (2):127–33.

Hansen, J., L. Holm, L. Frewer, P. Robinson, and P. Sandoe. 2003. "Beyond the knowledge deficit: recent research into lay and expert attitudes to food risks." Appetite 41 (2):111–21.

Kjærnes, U., M. Harvey, and A. Warde. 2007. Trust in Food. A Comparative and Institutional Analysis. London: Palgrave Macmillan.

Kjærnes, U., and R. Lavik. 2007. Farm Animal Welfare and Food Consumption Practices: Results from Surveys in Seven Countries. In: U. Kjærnes, M. Miele and J. Roex (eds). Attitudes of Consumers, Retailers and Producers to Farm Animal Welfare. Welfare Quality Reports No.2. Cardiff: Cardiff University

Lyon, J. D. 1998. "Coordinated food systems and accountability mechanisms for food safety: A law and economics approach." *Food And Drug Law Journal* 53:729–76.

Mishler, W., and R. Rose. 2001. "What are the origins of political trust? Testing institutional and cultural theories in post-communist societies." *Comparative Political Studies* 34 (1): 30–62.

Seligman, A. B. 1997. *The problem of trust*. Princeton: Princeton University Press.

Thompson, E. P. 1971. The moral economiy of the English crowd in the eighteenth century. *Past and Present* (Feb.), 76–136.

Tilly, C. 1975. Food supply and public order in modern Europe. In: C. Tilly (ed.) *The Formation Of National States in Western Europe*. Princeton, NJ: Princeton University Press.

Tilly, C. 2004. Trust and rule. *Theory and Society* 33, 1–30.

Trentmann, F., and F. Just. 2006. *Food and Conflict in Europe in the Age of the Two World Wars*. Houndmills: Palgrave Macmillan.

THE IMPORTANCE OF LOCAL PRODUCTION OF FOOD IN CRISIS SITUATIONS

ALENKA URBANČIČ

*National Assembly, Šubičeva 4,
SI-1000 Ljubljana, Slovenia
e-mail: alenka.urbancic@dz-rs.si*

Abstract: Food is an increasingly important strategic raw material. Due to climate change and extreme whether conditions in recent years we have been facing a critical shortage in crop production that together with an increased demand for food has resulted in the increased price of food. The need for cost reductions, increased need for food and access to food in stores have also caused important changes in the food production sector. The mitigation and adaptation to climate change is crucial. The local production of food is the appropriate system for implementing sustainability and food security and can reduce the lack of food and insecurity in a crisis situation. In general, the problems of food can not be solved without taking ethics into account. The ethical principles are the most easily introduced in the local production of food.

Keywords: Climate change, crisis situation, environment, ethics in food production, food security, local food production, sustainability

1. Introduction

It was not long ago that we all believed globalisation would bring benefits and welfare for most of the population of the world. We were all excited buying exotic food and seasonal products all year long, receiving information from other parts of the world in seconds, travelling easily around the globe. In fact, internationalisation is producing positive effects in some states, but it is also generating many negative results in others.

The leading role in this transnational flow of goods, services, capital and people, plays an elite group of rich countries, who are in league with international financial organizations and corporations. This is the reason for the strong perception that globalisation is a process driven only by capital without taking into consideration all the consequences. According to Noam Chomsky, one of the leading modern philosophers, the word globalisation is used to describe the neoliberal form of economic globalisation. In fact, free trade at first sight reduces prices but then eliminates social contract between states and societies and plays a negative role in the context of sustainability. Particularly in less developed countries the interactions of the local and the global have serious economic, food, health, community, and other security concerns.

However, global climate changes and a growing world population have also touched the invulnerable rich part of the World. Globalisation can not be partial.

2. Food provision in global context

Food provision is a very complex issue determined by many interactive factors. Generally, it depends on production, availability and access to food. Access to food is a function of economic and physiological potential and of food availability, which depends on production and distribution. Food production is a function of yield per unit area and the area from which the harvest is taken.

Growing population and climate changes are by far the most influential factors for food provision.

2.1. GLOBAL ENVIRONMENTAL CHANGE

Global environmental change is a term, which includes land cover, atmospheric composition, climate variability and climate, water availability and quality, nitrogen availability and cycling, biodiversity, and sea level.

Today there is no doubt that environmental change is the result of human activities, mostly in burning fuels.

The international group of scientists joined in IPCC (Intergovernmental Panel on Climate Change) received the Nobel Prize for Peace in 2007 "for their efforts to build up and disseminate greater knowledge about man-made climate change and to lay the foundations for the measures that are needed to counteract such change". Their findings are among others the following[1]:

- Observational evidence from all continents and most oceans show that many natural systems are being affected by regional climate changes, particularly temperature increases.

- A global assessment of data since 1970 has shown it is likely that anthropogenic warming has had a discernible influence on many physical and biological systems.

- Some large-scale climate events have the potential to cause very large impacts, especially after the twenty-first century.
- Impacts of climate change will vary regionally but, aggregated and discounted to the present, they are very likely to impose net annual costs, which will increase over time as global temperatures increase.
- Adaptation will be necessary to address impacts resulting from the warming which is already unavoidable due to past emissions.
- Vulnerability to climate change can be exacerbated by the presence of other stresses.
- Sustainable development can reduce vulnerability to climate change, and climate change could impede nations' abilities to achieve sustainable development pathways.
- Many impacts can be avoided, reduced or delayed by mitigation.

The increase in global average temperatures is very likely due to the anthropogenic concentrations of greenhouse gases (GHGs). All scenarios for future climate change in particular sectors predict[2]:

- Water system: increased water availability in moist tropics and high latitudes, decreasing water availability and increasing drought in mid-latitudes and semi-arid low latitudes, hundreds of millions of people exposed to increased water stress
- Ecosystem: great risk of extinction (even 30% or more), coral bleaching, increasing species range shifts and wildfire risk, ecosystem changes due to weakening of the meridional overturning circulation
- Food: complex, localised negative impacts on small holders, subsistence farmers and fishers; tendencies for cereal productivity to decrease in low latitudes and possibly at mid- to high latitudes or in some regions
- Coast: Increased damage from floods and storms, loss of wetlands, millions more people could experience coastal flooding each year
- Health: Increasing burden from malnutrition, diarrhoeal, cardio-respiratory and infectious diseases; Increased morbidity and mortality from heat waves, floods and droughts; changed distribution of some disease vectors

It is true that climate change affects much more less developed areas but also within areas with high incomes, some people (such as the poor, young children and the elderly) and some activities can be particularly at risk.

According ICCP findings[3] the matter of greatest concern should be:

- Water in the dry tropics
- Agriculture in low latitudes

- Human health in poor countries where activities depend on sensitive ecosystems, especially: tundra, boreal, mountains; or ecosystems already stressed: e.g. mangroves, coral reefs.

2.1.1. Scenario for European agricultural sector

It is clear that climate change has an exceptional influence on food provision. The production of food in Europe will change as follows[2]:

- Risk of inland flash floods and more frequent coastal flooding and increased erosion (due to storminess and sea level rise).

- In southern Europe, climate change is projected to worsen conditions (high temperatures and drought) in a region already vulnerable to climate variability, and to reduce water availability, hydropower potential, summer tourism and, in general, crop productivity.

All these changes will affect increased yields in colder environments, decreased yields in warmer environments and reduced yields in warmer regions due to heat stress. Large precipitation events will increase the damaging of crops, soil erosion and, in some areas, there will be impossible to cultivate land due to water logging of soils. In areas affected by drought will increase land degradation, yields will be lower due to crop damage and failure, livestock deaths will increase. Due to increased incidence of extreme high sea level salinisation of irrigation, water and fresh water system will increase.

On the other hand, agricultural practices, especially industrial agriculture, also contribute to climate change.

2.2. GROWING POPULATION

Projected increases in human population (Table 6.1) indicate that current production of food will need to be raised substantially over the next few decades.

TABLE 6.1. Population prospects 1950–2050 in billions (http://esa.un.org/unpp)[4].

	1950	1970	1990	2010	2030	2050
World	2.5	3.6	5.2	6.9	8.3	9.1
Europe	0.55	0.66	0.72	0.73	0.71	0.66
Africa	0.22	0.36	0.64	1.03	1.51	2.0
China	0. 55	0.83	1.15	1.35	1.46	1.45
India	0.37	0.55	0.86	1.22	1.50	1.66

In addition, large part of the population will live in the cities. It is necessary to take into account the improved diet of the new urban population in developing countries. Rising per capita incomes of this urbanised population will result in significant increases in the demand for food crops, meat, fish and forest products.

2.3. MISUSE OF LAND AND BIO FUEL PRODUCTION

Growing population and increasing development demand more and more agricultural land for other purposes: urbanization, infrastructure and production of raw materials for bio fuel.

Bio fuel is derived from biomass and is used for the production of heat or energy as an alternative environmental friendly fuel. Its production is driven by the global oil market crises. Because biomass used in processing bio fuel includes the agricultural products such as wheat, maize, soybean, rapeseed, sugar beet, sugarcane, palm oil etc., there is no doubt that bio fuel production affects food security. On the other hand this gives opportunities for farmers, especially in the developing world. Because of its ambiguous nature, production of bio fuels is on the Agenda of many meetings of the United Nations. FAO in its latest *State of Food and Agriculture 2008*[5] underlines that the growing demand for agricultural commodities for the production of biofuels is having significant repercussions in agricultural markets, and concerns are mounting over their negative impact on the food security of millions of people across the world. At the same time, the environmental impacts of biofuels are also coming under closer scrutiny. It is clear that even the use of bio-fuel decreases the emission of greenhouse gas, alot of energy is used for its production and plantations for bio fuels threaten biodiversity, forests and wildlife.

3. Local production of food as an answer to global crisis

As Khor[6] reported, many countries that were net exporters or self-sufficient in many food crops experienced a decline in local production and a rise in imports which had become cheaper because of the tariff reduction. Some of the imports are from developed countries, which heavily subsidize their food products. The local farmers' produce was subjected to unfair competition, and in many cases could not survive. The effects on farm incomes, on human welfare, on national food production and food security were severe. The current global crisis of high food prices, and of shortages in some countries, has given attention once again to food security concerns. In recent years there was complacency about food security and national self-sufficiency, as it was thought that cheaper imports would be always or usually available, and local food production was not so necessary as previously thought. Many developing countries reduced food production, many of them under advice of the international financial institutions. In crises, these countries have become very vulnerable and famine is their continuous threat.

Even world cereal production, this year, is expected to be large enough to meet anticipated utilization in the short-run due to favourable weather conditions, according to a FAO Report[7], experts warned that the current financial crisis will negatively affect agricultural sectors in many countries.

Local production has different meaning in different countries. The perception depends on the level of development. In less developed countries local production is a question of surviving but in rich countries the idea of local production is strongly connected with sustainability, organic farming, ecology etc.

Local production – is mainly meant to meet a high degree of self-sufficiency in the region. But on the other hand it is also important to encourage farmers and other owners of greater or smaller pieces of land to produce more varieties of crops for domestic use. It is not only important in the case of crisis situations, this is the way to turn thinking towards sustainability. Smaller part-time farms have also another important role, especially in countries without large flat areas: they are the key factor for preserving the cultural landscape.

Anyway, because of the new situation in the food market, the idea of "food security" has turned back to the traditional concept of greater self-sufficiency, instead of prioritising the option of relying on cheaper imports. It is now recognized that in the immediate period, there is need for emergency food supplies to affected countries, but that a long-term solution must include increased local food production.

Sustainable food production, processing, distribution, and consumption are guarantees for economic, environmental and social health of a particular region or state. The concept of local economies proclaims buying locally produced goods and services. It means simply to buy food (or any good or service) produced, grown, or raised as close to the home as possible.

More and more people have recognized that with industrialization, our food is now grown and processed in fewer and fewer locations and it has to travel further to reach the average consumer's refrigerator. The dominance of supermarkets in food retailing contributes massively to our dependence and vulnerability For example 80% of food eaten in London is imported. Although this is considered efficient and economically profitable for large agribusiness corporations, it is harmful to the environment, consumers and rural communities. There is also another threat: for capitals such as London, a food shortage would clearly be disastrous.

3.1. THE CASE OF INDIGENOUS PEOPLE FROM USA[8]

The Traditional Native American Farmers Association (TNAFA) has been working since 1992 to revitalize traditional agriculture towards building healthy communities and people. As a result of their efforts, there has been an increased interest in agriculture in the communities, and among the youth. TNAFA promotes family oriented farming as the best approach to develop agriculture, that ensures economic, social and health stability in their communities.

TNAFA's holistic approach employs agriculture as the starting point for building cultural pride, physical health, economic stability and ecological sustainable communities. They work with all members of the community's young, old, male, and female. One of their programs is the Traditional Agriculture and Permaculture Design Course on sustainable community design, which teaches youth participants such skills as building straw bale gabions, methods for assessing water quality, various planting techniques and garden designs, seed conservation techniques, preparation of traditional foods, cob building methods, harvesting of grey water and rain water, and using the dead pinion trees, which were victims of a 9-year drought, for making charcoal.

These "old fashioned" practises seem to be interesting from ethnical and traditional points of view but in a crisis situation could be of vital importance. It is clear that such way of life can't be a solution for a worldwide crisis but can offer an opportunity to think about some values and skills that have been lost with modernisation.

4. The Advantages of Local Production of Food in a Crisis Situation

4.1. LARGE SCALE OF CRISIS SITUATION

Practically at all times a great part of the world population is facing more or less a large crisis:

- Economic crisis resulting in low income and relatively high prices of goods;
- Poor, too rich, inadequate or/and not healthy diet resulting in inadequate nutritional status or even diseases
- Climate change crisis with frequent weather phenomenon such as hail storms, floods, extreme temperature, water shortage
- Natural disasters such as earthquakes, large floods, droughts and hurricanes
- Epidemical animal diseases such a BSE, "bird flu" etc.
- Wars, conflicts and terrorist attacks
- Psychological uncertainty and dependency

4.2. THE ADVANTAGES OF LOCAL PRODUCTION

Local production cannot be a solution for all emerging problems in such situations but could contribute a lot to better status of affected population. Local production for high degree of self-sufficiency requires a lot of knowledge about agriculture and about technologies for conservation and preparing food. This is extremely important in crisis when access to food stores is blocked. Nowadays, especially in more developed countries, we are facing a lack of knowledge for basic production, food storage and simple technologies. The majority of population could not survive or hardly survive for a long time without food supply. The farmers are market oriented and produce what is profitable, mostly monocultures to lower their costs. This fact makes farmers much more vulnerable than before. They depend on modern technologies that they are not able to cope without the help of special experts and they mostly have to buy the majority of food products. Local production helps to reduce vulnerability of the farmers and of the other population in the region.

4.2.1. Economic crisis

Local domestic production of seasonal food can significantly reduce the costs for every day life. Poor people spend a large part of their income for food and in addition to that, they are usually dependent on less quality food from big international supermarkets. Self-sufficiency of the most important food items on regional or state levels is one of the best tools to cope with economic crises in the field of food provision. The idea that globalisation is the reason for the world food crisis has been brought up. There are more and more articles blaming international organizations which promote free trade, that they destroyed self-sufficient economies in less developed countries and made these societies vulnerable and dependent.

4.2.2. Healthy diet

By choosing to eat lower on the food chain, and focusing on local and organic produce, we can avoid many toxic pesticides and enjoy fresh, tasty food. It is obvious that long transportation needs much more chemical protection than locally produced food. It is also easier to supervise the food producers and avoid large-scale infections, poisoning or diseases. Choosing organic means not only eating less treated food but also eating much more fruits and vegetables, which is very important for healthier diet.

4.2.3. Environment

Humans have been farming for hundreds of generations and for most of that time the production and consumption of food were highly connected with the cultural and social system. The production was integrated into natural circle balanced on the local level and, consequently, worldwide. Yet over just the last two or three generations, we have developed hugely successful agricultural systems based on industrial principles. They certainly produce more food per hectare and per worker than ever before, but only look so efficient if we ignore the harmful side-effects – the loss of soils, the damage to biodiversity, the pollution of water, the harm to human health.[9]

It is much easier to introduce sustainable agricultural practises on the local level. In fact, only the traditional way of farming can be called agriculture. The intensive farming with the utilization of huge amounts of mineral manures and pesticides can't avoid an impact on environment. Food for the local market can be produced without much of the fertilizers, pesticides and heavy machinery – characteristic of large – scale, single crop agro system. Sustainable farming is nature friendly.

The other important fact is less transportation which means less pollution or less GHGs caused by burning of mineral fuels and less waste of packing materials.

4.2.4. Psychological security

Human insecurity broadly conceived affects not just economic security but other areas of existence as well. The first major reference to human security in 1994 identified seven areas of concern and among others is also food security or access to food.[10] For most of human history, people have lived in a close relationship with the land and food has been closely connected to cultural and social systems. Foods had a special significance and meaning, as do the fields, grasslands, forests, rivers and seas yet now, for the first time, more people are living in urban rather than rural areas and they mostly depend on food bought in stores. If there is a threat of war or natural disaster, the first reaction of the population is to go to the stores and buy commodities, among them the most important is food. If there were a danger that food reserves would run out, people react in a panic and buy as much as possible. The scare and insecurity about food provisions have very negative effects on psychological status both on each person and on the population in general. The reactions are unpredictable and it is much more difficult to keep the population in a condition of reasonable reactions. With local production of food and with a certain degree of self-sufficiency it is possible to avoid total dependence on food bought in stores and this kind of security in crisis situations is very important.

4.2.5. Permanent knowledge and education

Local production and conservation of food requires a lot of knowledge. Especially ecological or sustainable agriculture are not possible without permanent education. The people who are aware of the importance of education and who are used to going to different courses are also more reliable and in the case of catastrophe much easier to organize them for immediate and efficient actions.

4.2.6. Ethics

Many problems we focus on today are the results of non-compliance with the principles of ethics. Over exploitation of natural resources and labor force, the use of non-reliable or even forbidden methods and additives for food or fodder to increase profit are some of the evident cases. The local production is the right field to introduce ethical principles. In fact, sustainable agriculture takes into account more than one ethical principle: respect towards the nature. But ethics is multidimensional and it could hardly be isolated. The people who learn how to respect the environment and all kinds of animals are expected to also be more sensitive to the problems of other people and prepared to help. This is very important fact in crisis situation.

5. General positive effects of local production of food

5.1. ENVIRONMENT

Trucking, shipping and flying food from around the country and the globe is harmful for the environment, for the climate and for public health. The investigation shows[11] that the average American foodstuff travels an estimated 1,500 miles before being consumed. Intensive farming contributes a lot to greenhouse –gasses and the solution for the environment is to introduce sustainable concepts and practises in agriculture. It is obvious that this process should start on the local level. Conventional agriculture follows only two goals: higher yields in higher prices on the market. To pursue these goals many practises have been developed without regard for their unintended consequences in a long-term period and without consideration of the environmental impact. Intensive tillage, monoculture, irrigation, application of inorganic fertilizer, chemical pest control and genetic modification are the separate methods used for higher productivity but each depends on the others and reinforces the necessity of using the others (Gliessman).[13]

Any transport means environmental impact with CO_2. In Austria they calculated that 10% more foods bought on local farms on the count of imported ones increased the domestic gross product by 1.5 milliards EUR and provided 17,000 jobs. The purchase of products from local farms contributes to better environment and climate as well as to economy[12]

(Weinberger). In the January issue of journal Ökoenergie the president of Austrian and European Eco-social Forum and ex-commissioner for agriculture in EU Dr. Franz Fischler wrote that we should think about using home produced foods as an investment into more sustainable future. In the same journal he and Dr. Helga Kromp-Kolb, professor at Agricultural University in Vienna, and Mr. Garhard Wlodkowski, the president of Agricultural and Forestry Chamber of Austria, stated that the purchase of home produced foods with short distance deliveries provided 450,000 jobs to Austrians and protected climate as well.[14] It was in the UK in 1990 that the idea of "food miles" came out. "Food miles" refer to the distance a food item travels from the farm to a consumer's home. The food miles for items bought in the grocery store tend to be 27 times higher than the food miles for goods bought from local sources.[11]

5.2. POSSIBILITY TO INTRODUCE ORGANIC AGRICULTURE

Organic farming is a form of agriculture that relies on crop rotation, green manure, stable manure, compost, biological pest control, and mechanical cultivation to maintain soil productivity and control pests, excluding or strictly limiting the use of synthetic fertilizers and synthetic pesticides, plant growth regulators, livestock feed additives, and genetically modified organisms. Organic production is a sustainable way of production and helps recovering the environmental damages.

5.3. BIODIVERSITY

Local sustainable farming helps a lot to keep biodiversity of plants, animals and micro-organisms. In recent years biodiversity has declined, some analysis predict the extinction of 30% of living organism by 2050.[1] Locally adapted plant and animal breeds are also more appropriate to local ecosystems and agricultural genetic diversity is a basic insurance against crop and livestock disease outbreaks.

6. Conclusions

1. Climate change, growing population and many crisis situations have great impact on current agricultural practises, which should be altered into more sustainable way.

2. Local production of food is long-term condition for sustainable agriculture and food provision.

3. Local production of food could significantly increase food provision and psychological security in a crisis situation.

4. Local production of food has significant impact on food safety and quality.

5. Local production of food can't be the only way for avoiding world famine but it is a good model for sustainable production and consumption of food.

6. Local production of food has many others positive effects on the environment and socio-economic status of the local community.

7. Governments and international organizations have to introduce and support policies – including educational – which strengthen rural people's capacity to cope with and mitigate climate change and face other emergency situations.

Bibliography

1. Parry, M.L., O.F. Canziani, J.P. Palutikof and Co-authors 2007: Technical Summary. Climate Change 2007: Impacts, Adaptation and Vulnerability. Contribution of Working Group II to the Fourth Assessment Report of the Intergovernmental Panel on Climate Change, M.L.Parry, O.F. Canziani, J.P. Palutikof, P.J. van der Linden and C.E. Hanson, Eds., Cambridge University Press, Cambridge, UK, 23–78

2. Climate Change 2007: Synthesis Report Summary for Policymakers. An Assessment of the Intergovernmental Panel on Climate Change (http://www.ipcc.ch/press/index.htm)

3. Watson, Bob. Presentation on the WG 2. http://www.ipcc.ch/graphics/presentations.htm

4. http://esa.un.org/unpp

5. FAO: The State of Food and Agriculture 2008. http://www.fao.org/sof/sofa/

6. Khor, M. Food Crisis, Climate Change and the Importance of Sustainable Agriculture. http://www.twnside.org.sg/title2/par/mk.3jun08.on.Food.Crisis.and.Climate. Change.rev1.doc

7. FAO News room. http://www.fao.org/news/story/en/item/8271/icode/

8. Contribution by Indigenous People. Advanced unedited text. Economic and Social Council of the United Nations, Sixteenth session, 5–16 May 2008

9. Pretty, J. Sustainable and Ecological Agriculture. http://www.essex.ac.uk/bs/staff/pretty/ sus_&_eco_agri.shtm

10. United Nations Development Programme, 1994:32

11. Fossils fuel and energy use. http://www.sustainabletable.org/issues/energy/

12. Gliessman, S. Engles, E., Krieger, R. Agro ecology: Ecological Processes in Sustainable Agriculture. CRC Press, 1998, pp. 1–3

13. Weinberger, K. Kurze Transportwege schützen das Klima. Ökoenergie, 57, Dezember 2004/Jänner 2005, p. 16

14. Osterc, J. Sustainable Cattle Production is the Best for Slovenia. Animal Science Days, Strunjan, 17–21 Sept. 2008, 8 p

TERRORISM "ASSESSING THE THREAT FOR CRITICAL INFRASTRUCTURE" METHODOLOGIES FOR WATER SUPPLIER

DI WOLFGANG CZERNI

Geschäftsführer der Infraprotect GmbHGesellschaft für Risikoanalyse, Notfall- und Krisenmanagement Wolfsaugasse 11/6/13, 1200 vienna,
e-mail: w.czerni@infraprotect.at

Abstract: Operators of water supply infrastructures will find tons of information on terrorism against critical infrastructures. In the close history of middle Europe, we can come up with many criminal attacks against water suppliers. The attack against the "Bodenseewasserversorgung" in fall 2005 demonstrated again, that water supply is clearly a target for criminals and/or terrorist. Before this event, specially after this repeated event then, many efforts where made to secure the supply chain for providing pure drinking water to our citizens. Infraprotect itself was and still is involved in many research programs to secure drinking water against terrorism. The aim of all these programs beginning by the EPCIP[1] program down to national programs in various countries is to identify and quantify the threat to water supply. In fact there are two common approaches to identify threats; the most common used approach is risk based followed by an effect based approach only.

This work describes and compares the methodologies of water supply operators and the concerns of military leaders, securing water supply for citizens during peace keeping operations.

Keywords: Water supply threats, fwater security, critical infrastructure protection, terrorism, risk assesment

1. Framework and general conditions

For military leaders the operational requirements for water supply of personnel are clearly defined and will not be further touched here. The civil–military

---

[1]EPCIP, European program critical infrastructure protection

V. Koukouliou et al. (eds.), Threats to Food and Water Chain Infrastructure,
DOI 10.1007/978-90-481-3546-2_7. © Springer Science + Business Media B.V. 2010

cooperation needs in a transition state of combating and peace keeping state with a defined asymmetric threat or undefined asymmetric threats aren't really subject of detailed discussion yet. It's remarkable, that threat- and risk assessment methodologies for assessing terrorist threats are clearly driven by military needs. -So far-. For the operators of water supply infrastructures within middle Europe, the security focus is more ore less securing against criminal attacks. Initial question such as:

- Technical based threat scenarios
- Vulnerability and target analysis for exploitation, production, storage, distribution or retrieval of untreated water
- Threat driven preventional measures
- Aren't necessarily guided through a legal framework
- In fact, coping with terrorist threats against water supply isn't comprehensive covered by water supply operators

2. Basics

In comparison to military operations, threat and risk assessment assets of critical infrastructure provider are focused and tailored to daily business needs. The realistic threat spectrum itself can be summarized to:

- Conventional threats (e.g., bombs against pipes, or buildings)
- Conventional threats against supporting processes (electricity, wiring of the process control systems
- Unconventional threats (poisons or radioactive materials) against water itself, personnel or materiel

A quick characterization of water suppliers "production sites" leads to the statement, that, their sites are usually well known. The knowledge of the operating personal is often very limited to technical means, such as mass flows, operating hours of pumps, storage volumes, etc. Unfortunately many water suppliers aren't aware of their vulnerabilities to terrorism.

3. Risk- and threat assessment methodologies

According to the ACO Directive 80-25 Force Protection the Risk- and Threat – Assessment process has four basic steps (Figure 7.1):

- Identify your risks
- Assess the risk

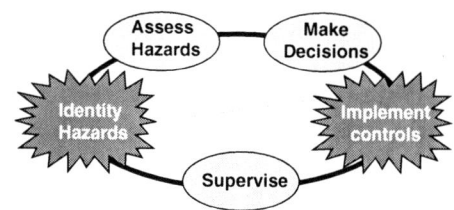

Figure 7.1. Force protection steps.

- Develop Risk Management Plans and Make a Decision
- Implement risk management plans

These four steps, assume that the organization behind has a minimum of readiness in order to cope with emergency situations. We have learned in the past, that water supplier aren't trained to cope with emergency situations. The deal mainly with individual technical problems and do not have the knowledge to cope with complex situation, such as suicide attacks against personnel or CBRN attacks against storage or distributions sites.

Therefore one of the lessons learned in the past was, that the risk- and threat assessment process has to consider the lack of knowledge coping complex situations within those organisations.

An adopted risk- and threat assessment process had to be designed. Considering this, a very simple but robust solution process will be presented consisting of five phases:

- Planning within the organisation
- Risk analysis process adopted to the needs of the specific organisation
- Preventive measures strategies
- Setup up of an holistic emergency and crisis management process
- Evaluation process, evaluating the risks and the management of the organisation

4. Planning needs within the organisation

In an initial phase, fundamental questions are to be raised:

- Responsibilities are anchored where in the organisation? Especially the definition of responsibilities for implementation are to be defined
- Availability of resources (personnel, finance, material)
- Legal obligations to establish risk and crisis management
- Last but not least strategic protection aims are to be highlighted by the management board

5. Adopted risk analysis process

It is very important to understand, that this is the fundamental process within the organisation in order to draw appropriate conclusion, especially for coping with extremely complex situations. In fact the risk analysis process shall provide a structured overview of possible threats to production and distribution processes and their vulnerabilities implemented by the organisation.

Combining and assembling all threats yield a realistic picture for all critical processes in the organisation for individual scenarios. The aim of these steps is, to provide an easy-to-understand picture of what can easily happen, which scenarios are technically likely to be used by terrorist, in order to gain a feeling on realistic dimensions of the effects to personnel, material and environment.

The adopted process starts by splitting the organization into specific production, distribution, storage and support processes and sub-processes. The level of detail is decision to the VAT.[2]

6. Overview of the plannend process

Above figure describes the necessary steps for a comprehensive risk assessment (Figure 7.2 and 7.3).

7. Preventive measures strategies

Preventive measures help reducing identified risks to critical sub-processes. They help achieving operational protection aims (Figure 7.4).

One conclusion out of the experience of the past 15 years is, that water supplier have to decide where in the complex-coping triangle preventive measures should be subject to a cost-benefit analysis aiming reducing the overall risk.

This is done by comparing potential expenditures and the direct and indirect costs resulting to the organization from an extreme incident.

However, measures to reduce risks that are unlikely to occur but would have dramatic impacts if they happen are often impossible to justify on the basis of risk and cost-benefit analysis alone. In such cases, it may help to consider societal and ethical aspects as well as the legal framework conditions

---

[2]VAT Vulnerability Assessment Team

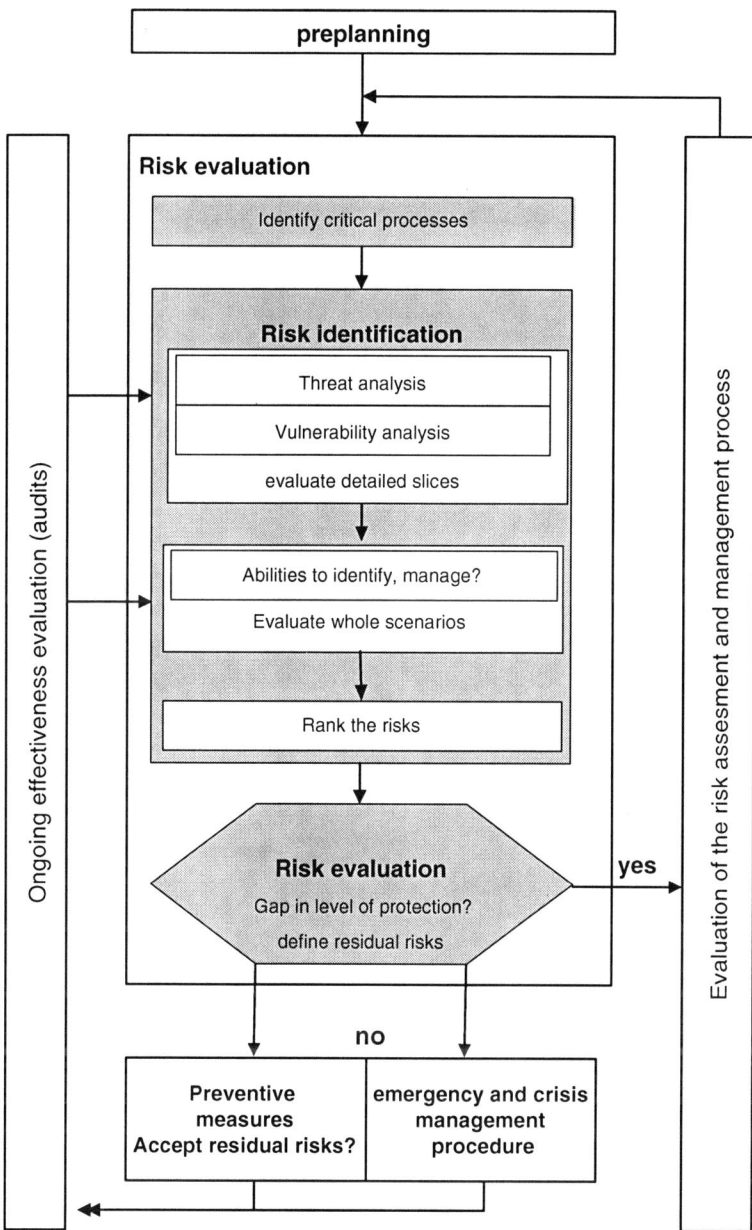

Figure 7.2. Risk assessment and management process.

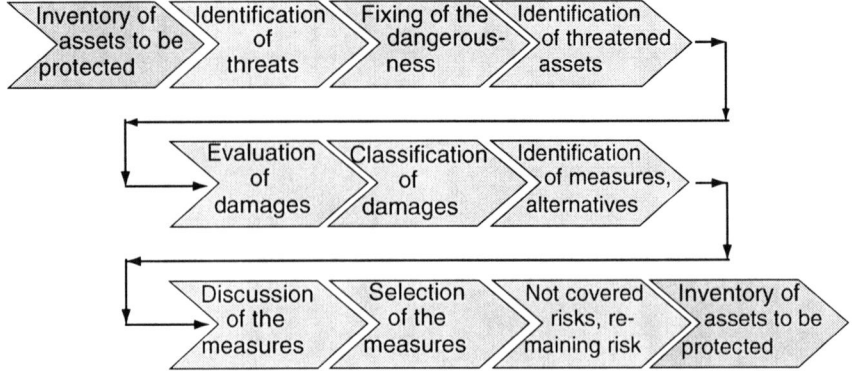

Figure 7.3. Risk assessment and management process.

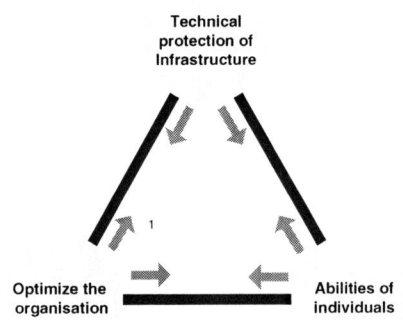

Figure 7.4. Risk assessment and management process.

when deciding on protective measures. Preventive strategies take advantage of risk avoidance and risk acceptance. They can only be used parallel with risk-reduction measures mainly based on optimizing physical risks.

8. Emergency and crisis management process

Successful crisis management bases on management strategies, such as risk management described above. Emergency and crisis management focus on preparing and activating measures to keep the organization running. It helps to ensure business continuity. The aim of the management process is, to return to normal operations as soon as possible. In the literature one will find tons of information on how to implement crisis management procedures.

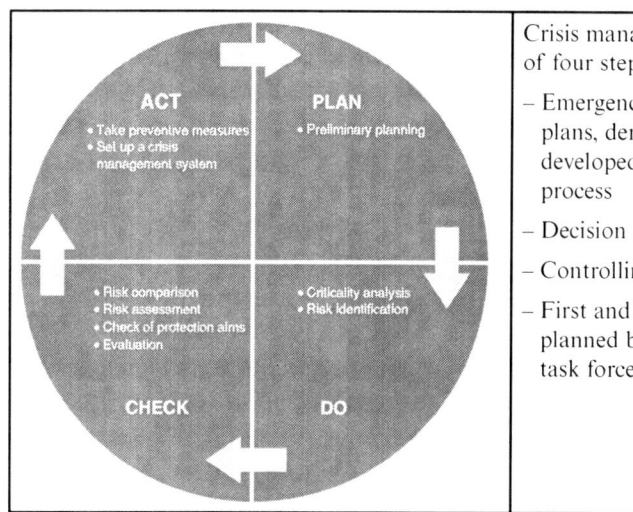

Crisis management touch a minimum of four steps:

– Emergency and crisis management plans, derived from the scenarios, developed by the risk-management process

– Decision making procedures

– Controlling procedures

– First and second responders actions, planned by the crisis management task force

Most water supplier don't have operational know how coping with CBRN agents. It's very important, that all first responders and the organisation attacked by terrorists/criminal have established communication procedures in order to optimize cooperation.

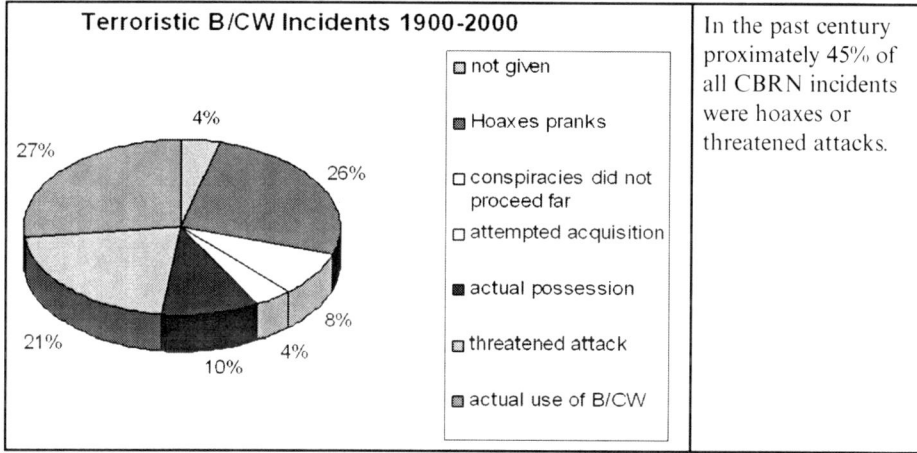

In the past century proximately 45% of all CBRN incidents were hoaxes or threatened attacks.

Although we have only 8 years of statistics behind us now in the new century, we can summarize, that the realistic threat scenario, based on numbers, has not dramatically changed yet.

It's obvious, that crisis management procedures cannot be derived from this knowledge. Derived from the given numbers of the last years, we can show the following tendencies:

– The total numbers of attacks will decrease.

– The casualty numbers per attack will increase.

One can dilute from that, that the attacks will be planned better and more efficient. Bringing this circumstance to our minds, we can point out, that the risk management process gains more importance. The intensity of work, once attacked, will also increase. From this point of view, we have to underline, that the crisis management procedures and all preparedness measures must be maximized to this circumstances.

9. Evaluation process

Most guidelines lack of a state-of the-art evaluation process to provide adequate lessons-identified and lessons learned procedures. For coping terrorist threats, evaluation of the own crisis management system during or after an incident of neighbouring water suppliers should be a key asset. Usually, most important tasks of crisis management are:

– Creating the conceptual, organizational and procedural conditions needed to deal with an extreme incident as effectively as possible.

– Establishing special structures to respond in case of crisis

– In particular setting up a crisis task force

Focusing on the lessons learned process, the following points are important to decision makers in order to clearly enhance the effectiveness of the own crisis management capability. This process must:

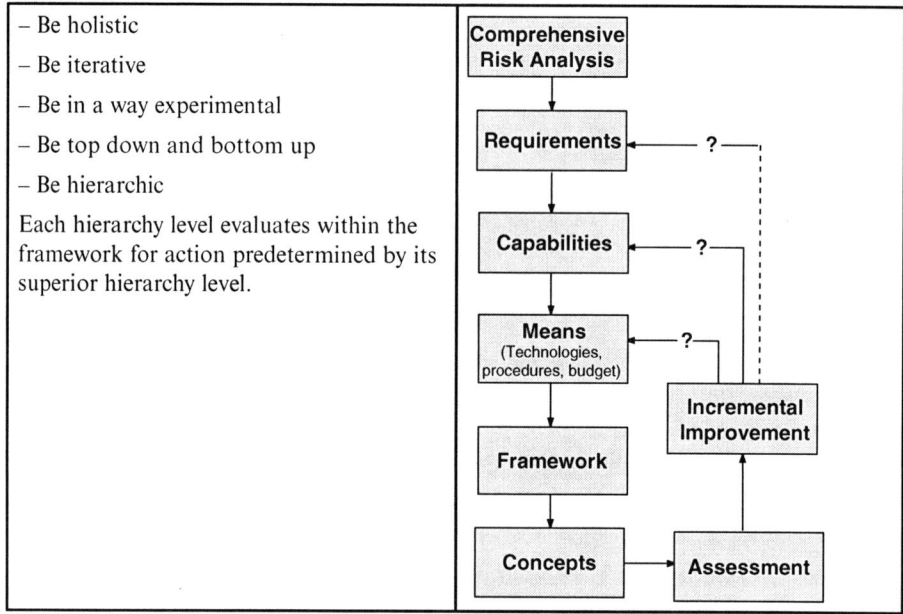

– Be holistic

– Be iterative

– Be in a way experimental

– Be top down and bottom up

– Be hierarchic

Each hierarchy level evaluates within the framework for action predetermined by its superior hierarchy level.

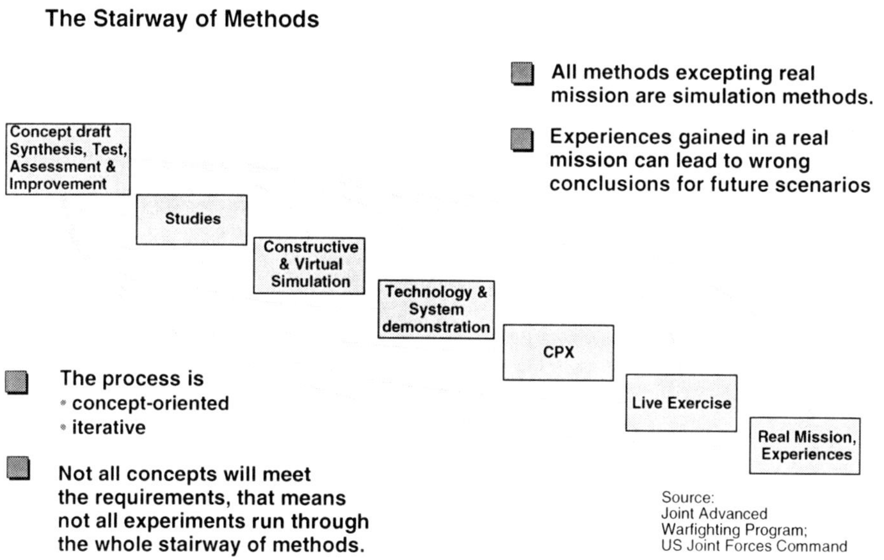

The Stairway of Methods

■ All methods excepting real mission are simulation methods.

■ Experiences gained in a real mission can lead to wrong conclusions for future scenarios

Concept draft Synthesis, Test, Assessment & Improvement

Studies

Constructive & Virtual Simulation

Technology & System demonstration

CPX

Live Exercise

Real Mission, Experiences

■ The process is
• concept-oriented
• iterative

■ Not all concepts will meet the requirements, that means not all experiments run through the whole stairway of methods.

Source:
Joint Advanced
Warfighting Program;
US Joint Forces Command

Figure 7.5. The stairway of methods.

The evaluation concept is a step by step method, derived from military mission planning expertise (effect based mission planning).

The stairway method is also a guideline for implementing an effective crisis management process (Figure 7.5).

The evaluation process covers all phases of risk and crisis management, from checking the items identified during preliminary planning, to checking whether risk profiles are current and whether preventive measures taken and the crisis management system are effective.

We have learned from the past, the terrorists adopt their attack procedures and learn from each other. Since terrorist attacks against water supplier happen rarely our evaluation process should be carried out regularly, preferable once a year. Especially after installing new physical security barriers, after expansion or restructuring of the organization additional evaluations are necessary.

Risk- and threat assessment and/or and crisis management must be taken seriously. Those measures can only provide the necessary level of security if all phases are regularly tested.

Bibliography

ACO Directive 80-25 Force Protection
Antiterrorism Vulnerability Assessment Team Guidelines, Dtra 2002
AS/NZS 4360:2004; Australian/New Zealand Standard Risk Management, Standards Australia, Standards New Zealand, Third Edition 2004

BMI, BKK, Protecting Critical Infrastructures. A guide for companies and government authorities Risk and Crisis Management

EPCIP European program critical infrastructure protection program

FwDV 100, Feuerwehr-Dienstvorschrift 100, Führung und Leitung im Einsatz, Stand: 10. März 1999

Gustin Joseph F., Disaster & Recovery Planning: A Guide for Facility Managers, Fairmont Press Inc

Haimes Yacov Y., Risk Modeling, Assessment, and Management, Wiley, Second Edition, 2004

Hirschmann, Kai: Das Phänomen "Terrorismus": Entwicklungen und neue Herausforderungen; in BAKS: Sicherheitspolitik in neuen Dimensionen; Verlag E.S. Mittler & Sohn GmbH, Hamburg Berlin Bonn, 2001

International Risk Governance Council, White paper on managing and reducing social vulnerabilities from coupled critical infrastructures, October 2006

Lewis, Ted G. Critical Infrastructure Protection in Homeland Security – Defending a Networked Nation, Wiley, 2006

Rosenthal, Uriel: Crisis management: on the thin line between success and failure. In: Asian Review of Public Administration, 1992

Schoch, Bruno: Der neue Terrorismus: Hintergründe und Handlungsfelder in arabischen Staaten; in Kai Hirschmann / Christian Leggemann (Hrsg.): Der Kampf gegen den Terrorismus: Strategien und Handlungserfordernisse in Deutschland; Berliner Wissenschafts-Verlag GmbH; 2003

The Business Continuity Institute, BCI, Business Continuity Management, Good Practice Richtlinien, 2005

SESSION II
MANAGEMENT

WATER MANAGEMENT IN AUSTRIA AND SECURITY OF WATER SUPPLY

WILFRIED SCHIMON

Federal Ministry of Agriculture, Forestry, Environment and Water Management, Marxergasse 2, 1030 Wien, Austria e-mail: wilfried.schimon@lebensministerium.at

Abstract: Water Management in Austria is based on a real abundance of water in great parts of the country. Related to the water availability per capita, Austria is ranked in the upper range in Europe. However, the distribution of precipitation and thus the availability of the resource is not even. In the eastern and south-eastern regions of the country, temporal problems with regard to water quantity can occur in dry years. To remedy these problems, interregional water supply systems were installed.

Comprehensive and substantial water policy activities guarantee the availability of water resources in terms of sufficient quality and quantity.

In Austria, domestic water supply is exclusively based on groundwater resources which usually do not require a prior removal of pollutants. The ground water abstracted includes water in porous geological formations as well as spring water from karst massifs. Drinking water can be extracted nearly everywhere in sufficient quality and quantity. This fact is leading to the existing structure of water supply in Austria, which is characterized by only few large but a high number of small water supply facilities. Domestic water supply is largely in the hands of the public sector and carried out by the municipalities.

Water supply must be considered – at least regionally – as a critical infrastructure. Potential threats for water supply can be various and include

- accidents, involving water-polluting shubstances which are transported on roads and railways
- contaminations of catchment areas from landfills or industrial facilities
- contaminations of catchment areas or parts of the water supply infrastructure as a result from natural hazards like floods
- destruction of pipelines and installations by floods or mudflows
- power blackout, failures or misuse of data transfers and telecontrol
- intentional attacks on water supply facilities with criminal or terroristic background
- contamination by unclear fallout caused by nuclear reactor accidents

V. Koukouliou et al. (eds.), Threats to Food and Water Chain Infrastructure,
DOI 10.1007/978-90-481-3546-2_8. © Springer Science+Business Media B.V. 2010

Of course, the mentioned potential threats show different probabilities of occurrence and different dimensions in the expected effects and impacts. Provisions by law only cover the precautions against possible conventional negative impacts on water quality and quantity. This is also reflected by the establishment of protected areas, which are aiming to prevent water from deterioration by such conventional threats. Drinking water protection zones in particular proved to be successful for the preservation of the quality of drinking water.

With regard to exceptional threats, no legal precautionary provisions are in place. However, the guideline "Drinking Water Emergency Supply" was elaborated in the framework of the Austrian Association for Gas and Water (OVGW), enacted in the year 1988 and revised in the year 2006. This guideline covers the entire spectrum of potential exceptional threats with tailor made measures. The implementation of those measures is within the direct responsibility of the operator of the water supply facility.

The guidelines proved to be successful in terms of water suppliers being prepared for worstcase scenarios. Planning was completed, the necessary equipment purchased and agreements with the emergency services concretised. In general, large water suppliers are usually prepared in a better way for exceptional scenarios compared to small facilities. This is caused by the larger financial potential and the availability of qualified human resources. Cooperation of water suppliers and sharing experiences can help to improve performance.

Emergency organizations, in particular specialized units of the federal army, NGOs and firefighters are of high importance in this context. These organisations are able to offer substantial support to water suppliers in case of exceptional and extreme situations. Furthermore, the federal army and the Red Cross have experience in successfully supporting the installation of autonomously operating water supply units at international missions. Therefore, special knowledge and understanding wih regard to this sector was gained by these organisations.

Based on that experience it was possible to provide Austria's population with drinking water meeting drinking water standards also in exceptional situations like the floods in the year 2002. Implementation of the measures, as listed in the "Drinking Water Emergency Directive", into practice has fully proven its efficiency in Austria. The emergency response in 2002 covered for instance the whole range of measures from the provision of packaged drinking water, the operating of mobile drinking water facilities up to the remediation of impacted installations in place.

Last but not least one important aspect has to be addressed – the permanent practical training and the periodical check and review of emergency operations at a larger scale. Especially the federal army and emergency organizations cover and deal with this issue. However, emergency exercises and co-operation with water supply enterprises take place very rerely on national but also on internaional level. There is the clear need to deal with the issue of emergency water supply within exercises and practical training at all levels and in all dimensions. One of the most important aims of training should be to raise awareness of water suppliers for emergency situations and to improve the co-operation of water supply operators with emergency organisations.

Keywords: Water management, Austria, water supply, threats, drinking water, emergency

1. Basic facts on water management

1.1. GLOBAL WATER MANAGEMENT

Water management including water protection and water supply has a strong connection with security aspects of water and preparedness against threats to drinking water.

At first I would like to have a glance at the global situation of fresh water resources. Our planet is known as the "Blue Planet". As two third of its surface are covered by oceans it appears in a deep blue colour, seen from the space.

But this appearance is misleading, concerning availability of water for human purposes. About 97% of the global water resources are salt water, 2% are fixed in glaciers and pole caps. There remains 1% fresh water, being part of the water circle and forming groundwater. Also availability of this 1% is restricted because of technical limitations of use and because of environmental contamination.

Precipitation is distributed quite unequal over the surface of our world and focuses of density and growth of population are not in congruence (Figure 8.1).

There are more than 300 river catchment areas in the world shared by several countries. All countries in a catchment area need to use the discharge of the river. In regions which are poor of water, the distribution of the discharge can create conflicts between the countries. Until now, no war was caused only by waterborne problems, but sometimes this problems contributed to their reasons. Movement of millions of people actually is originated by water problems. And nobody can exclude the possibility of wars because of water shortages in the future.

In the latest decades rules of international law have been developed for an equitable use of such transboundary waters and have proofed to be successful in many cases. The crucial point for a good cooperation is the institutional framework of a good working commission.

The greatest part of Austria lies in the Danube River Basin. Already in times, when the Iron Curtain divided Europe into two parts, Austria initiated a cooperation of the countries in the Danube region. The basis for it was the Declaration of Bucharest. In 1998 the first meeting of the International Commission for the Protection of Danube River (ICPDR) took place in Vienna (Figure 8.2).

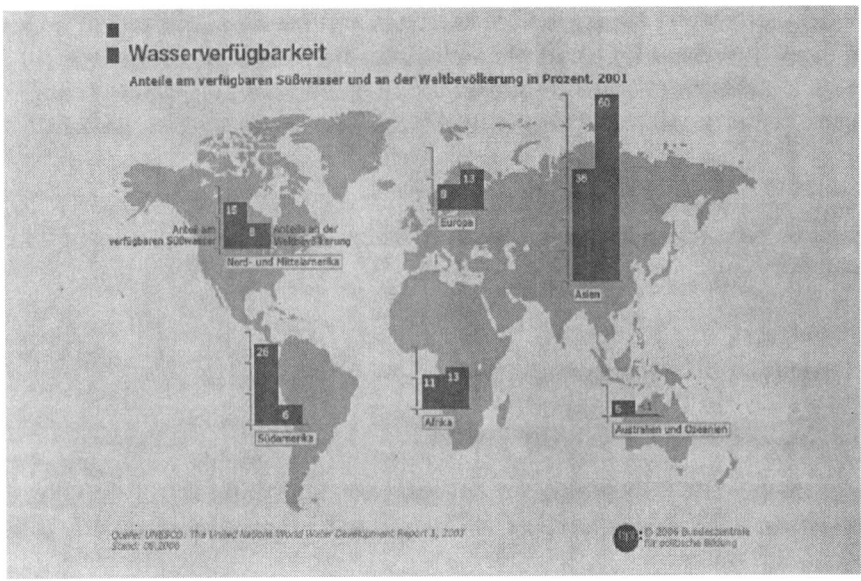

Figure 8.1. Water availability: Share of available fresh water (*blue*) and share of world's population (*green*).

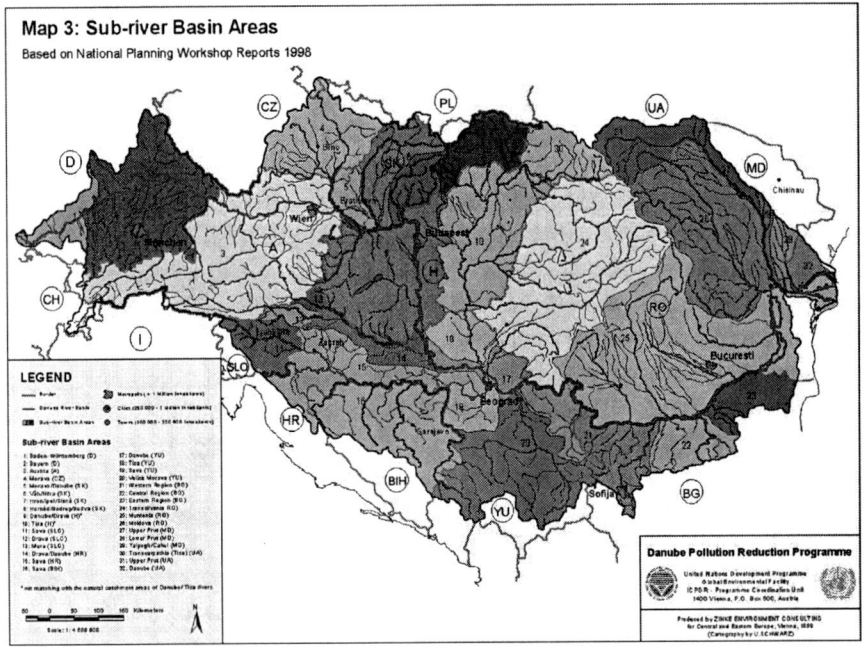

Figure 8.2. The Danube River Basin.

The ICPDR is regarded as an example of best practice for cooperation in a river basin. In the recent years the ICPDR got strongly committed to water related regulations of the EU, in particular to the EU-Water Framework Directive (WFD).

The EU-Water Framework Directive (WFD) came into force in the year 2000. It is effective far beyond Europe. It addresses the important global water Problems: Pollution of waters, destruction of habitats and overexploitation of groundwater. The main issue of the WFD is the demand for cooperation in transboundary river basins. Therefore the WFD as a new approach creating international cooperation for the equitable use of transboundary waters can be regarded as an important instrument of international security and of peace.

1.2. WATER MANAGEMENT IN AUSTRIA

The abundance of a country in water can be indicated by the water availability. That means usable water resources per capita of inhabitants. Austria lies in the upper European middle field concerning this characteristic value. Austria is rather independent from influx from other countries, as the main part of water resources is fed by precipitation. There are great differences in precipitation in the area of Austria, from 2,500 mm in the alpine regions to 600 mm in the eastern and southeast parts of Austria. In these regions at short notice quantity problems can occur in dry years. Interregional water transfer systems are important instruments for remedy of these problems (Figure 8.3).

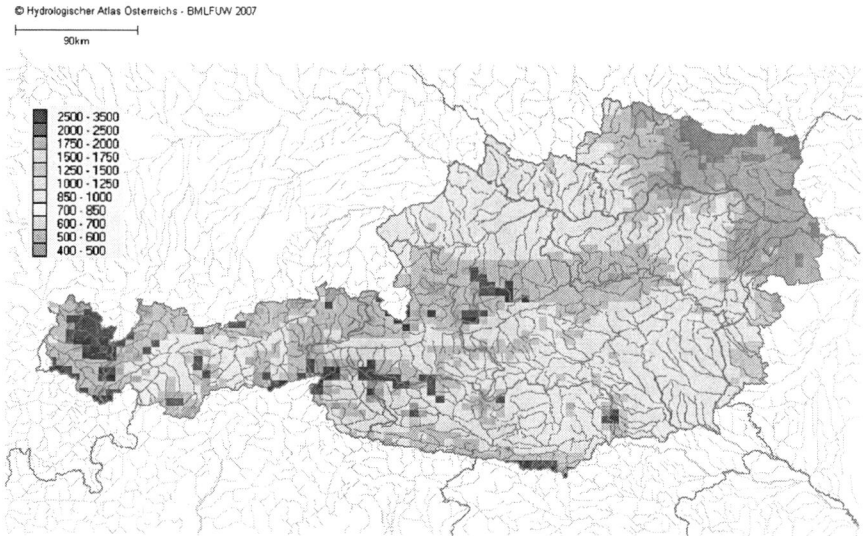

Figure 8.3. Distribution of total precipitation in Austria.

In Austria a consequent water protection policy has been performed over more than four decades. Starting in the 1960s this policy was focussed at first on the lakes, than on rivers and the groundwater.

For domestic water supply exclusively groundwater is used. Groundwater protection is an important precondition for it. It has two aspects – ground water quantity and quality.

The goal to use groundwater without treatment for drinking water is a basic principle of water management in Austria and gives groundwater protection high priority. In former years the pesticide Atracine caused some large scale contamination of groundwater, but as the application of this herbicide is forbidden in Austria this problem now is generally solved.

The most important problem is still nitrate, especially in the eastern parts of Austria, which are less rich in precipitation, than the alpine areas. A lot of instruments are applied to reduce impact of agricultural landuse, I only will mention the National Action Program according the EU Directive on nitrate. The development of nitrate in groundwater is a function of climate conditions as well as of agricultural management. The trends in agricultural practice like growth of cultures with energy plants possibly can increase the impact of nitrate to groundwater (Figure 8.4).

The protection of groundwater quantity aims at a balance of abstractions from groundwater and the regeneration of groundwater. The risk analysis performed in 2004 according to the EU Water Framework Directive showed

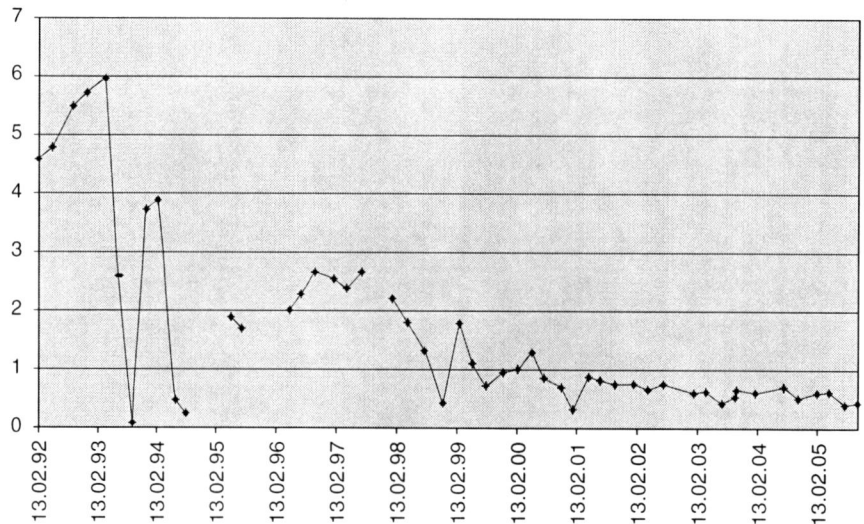

Figure 8.4. Reduction of Atracine (mg/l) at a typical monitoring point in Austria.

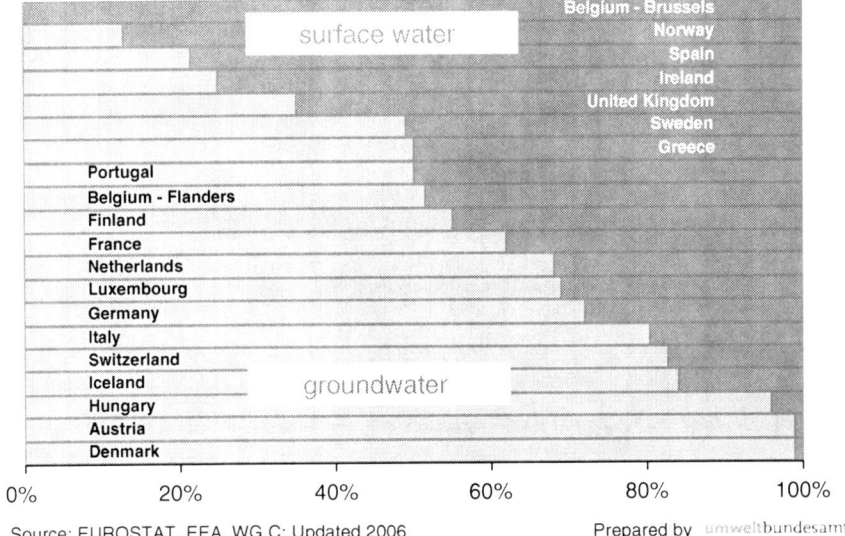

Share of ground and surface water in the public
water supply of Europe

Source: EUROSTAT, EEA, WG C; Updated 2006   Prepared by umweltbundesamt

Figure 8.5. Share of ground- and surfacewater in the domestic water supply in Europe.

that in Austria all groundwater bodies are in a good condition concerning quantity and until now there is no danger of overexploitation. In future climate change will possibly make situation worse in some regions of Austria.

In Europe most countries use treated surface water for domestic water supply. In Austria domestic, industrial and agricultural water supply is based only on groundwater. The term groundwater should be understood in such a way that it covers as well water in porous geological formations as spring water from the karst (Figure 8.5).

For purposes of domestic water supply about 700 Mio.m³/year are abstracted. Drinking water can be extracted nearly everywhere in Austria in sufficient quantity and quality, so even use for domestic water supply generally requires no conditioning. The result is a structure of water supply systems, which contains only few large but a multiplicity of small water suppliers. Only two enterprises provide more than 300,000 inhabitants with water. In Austria domestic water supply is largely in the hands of municipalities (Figure 8.6).

About 90% of the population is served by central water supply systems.

Key figures show that water consumption has been nearly unchanged since a lot of years. In the recent decades strategies were performed to stop water losses in the systems of domestic water supply.

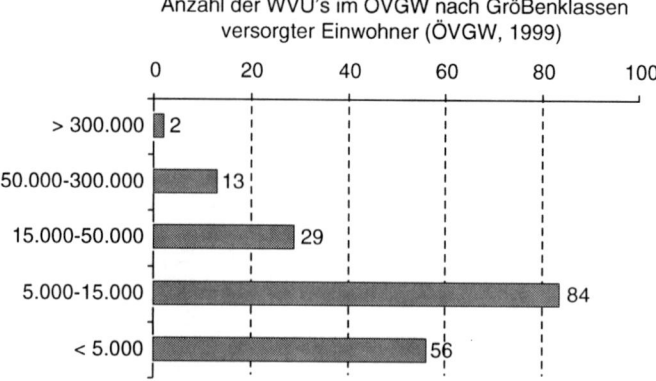

Figure 8.6. Number of water supply enterprises according supplied inhabitants (1999).

Industrial water demand for production is generally covered by groundwater, about 700 Mio m³/a of it are abstracted by the enterprises. For cooling purposes the enterprises generally use surface water (1,000 Mio m³/a, excluded cooling water for thermic power plants). The increasingly restrictive regulations for industrial waste water led to technologies more economical in water consumption.

Agricultural irrigation is applied in the eastern and south eastern parts of Austria. Generally farmers use groundwater, available near their fields. Therefore abstraction is decentralized and generally pumps are run by private fuel engines. The total amount of agricultural groundwater abstraction is estimated by about 200 Mio m³/a.

2. Threats to water supply

2.1. POTENTIAL THREATS AND REAL INCIDENTS

Systems of domestic water supply are vulnerable to numerous pressures caused by disasters and modes of attacks. The potential threats of the water supply can be various:

- Accidents with water-polluting material on roads or railways
- Contamination of catchment areas from landfills or factories
- Contamination of catchment areas or facilities of water supply as a result of floods
- Destruction of pipelines and installations by floods or mudflows
- Failures in electric power supply, failures or misuse of data transfers and telecontrol

- Intentional attacks on the facilities of water supply with criminal or terrorist background
- Contamination by fall out after a nuclear reactor accident

Examples proof, that threats of this kind are not only potential, but already occurred and so are very realistic:

- Large scale chemical contamination of the River Rhine after the burning of a chemical plant
- Heavy metal contamination of the River Danube as a consequence of accident-caused release of mining sludge
- Artillery fire on the few remaining stand pipes in Sarajevo during Balkan crises
- Attempted attack on an embassy in an European Capital, inserting poison into the water supply system
- Flood disasters in the years 1997, 2002, 2005, 2006 in a nearly continental scale
- Attempted attack with pesticides on a water treatment plant of a regional water supply at Lake Constance
- Accident of Chernobyl with nuclear fall out on great parts of Europe

The impacts can affect the abstraction area and the water resource as well as the facilities of water supply like pipeline networks, pumps, systems of telecontrol etc.

2.2. IMPACT ON ABSTRACTION AREAS AND WATER RESOURCES

The vulnerability of a water resource depends from its type and the type of aquifer. In the case of radioactive fall out, for example as a consequence of an accident in a nuclear power plant – groundwater in porous media is much better protected, than groundwater in karstic media.

A very significant experience concerning potential danger of contamination of water supply was the disaster of Chernobyl which happened on 26 April 1986.

One of the abstraction areas of Vienna water supply lies in Wien-Nußdorf, very near to the River Danube, the abstracted water is predominantly surface water, passing through about 6 m of sand and gravel. The average ß-activity of the water of the River Danube and the filtrate in the time before the accident of Chernobyl was nearly equal in the range of 0.1–0.3 Bq/l. In the time of May 1986 the ß-activity of water of the river reached a level of 8.1 Bq/l with maximum of 54 Bq/l, while the filtrate only reached an activity of 0.2 Bq/l in average with 0.5 Bq/l as maximum.

This experiences show that the passage through the filterbody of the banks, although it is not a perfect soil-passage, can keep the increase of activity on a low level, even in accordance with the drinking water limits valid at these times.

The I. and the II. Hochquellenwasserleitung (HQL) are the most important sources of drinking water for Vienna. They gather water from karstic areas in the mountains about 150 km south of Vienna. The ß-activity in the months before the accident of Chernobyl amounted less than 45 mBq/l. Because of the Chernobyl-fall out the ß-activity rose to 2 Bq/l in the I. HQL between 30 April and 3 May and to 3.8 Bq/l in the II. HQL between 4 and 6 May. Values of the normal level were reached in the I. HQL already after 8 May, in the II. HQL not before end of 1986. The comparison of the situation in the two catchment areas shows clearly the influence of different precipitation (<5 mm in the area of the I. and 10–15 mm in the area of the II. HQL), different altitude and therefore different temperature.

Summing up different types of water resources have different vulnerability to nuclear fall out:

- Groundwater in porous media, covering layer with clay and humus proved well protected.

- Filtration of surface water of Danube through few meters of sand and gravel reduced ß-activity in the filtrate with the factor 100 compared with the surface water.

- Groundwater in karstic areas can show very different behaviour depending from hydrogeological and climatic circumstances.

2.3. IMPACT ON THE FACILITIES OF WATER SUPPLY

The pipeline network, pump stations etc. can be vulnerable to demolition, by acts of sabotage as well as by natural disasters like floods. But repair work of this kind is a matter of routine and does not need a lot of time, which additionally in most cases can be bridged by using the content of reservoirs. So such attacks are not very dangerous for water supply.

An even more serious problem would arise with an application of NBC substances by an aggressor. Thereby we have to assume that the aggressor has a good knowledge in the choice of the substance and the point of attack.

Taken the case, we find hints for unauthorized manipulation on the facilities of water supply, which make an application of chemical substances seem likely or even probable – for control of water supply hereof a great problem is raising – the identification of an unknown substance – out of a great number of different possible substances. My ministry started to deal with this issue – and we have done the first step, an attempt of prioritisation of substances according to their relevant features for acts of sabotage. Now it is up to a next step to draw consequences from these results concerning recommendations for this specific analytical challenge.

In our opinion it would be advisable to create a "crisis lab" by cooperation of relevant labs with the intention of providing a broad spectrum of analytical methodical experience and capacities. 24/7 service for emergency situations is also an issue to be dealt with. With regard to organisation and financing of a 24/7 service the existing emergency services should be part of the cooperating network.

2.4. WATER SUPPLY AS CRITICAL INFRASTRUCTURE

Water supply has to be seen as an – at least regionally – relevant critical infrastructure. It is relevant not only from the technical point of view but also with regard to the emotional impact of a break down of water supply on the public (Figure 8.7).

On the other hand water supply is strongly depending from other infrastructures, so

- Energy, specially electric power
- Telecommunication as a part of control systems
- Transport facilities to bring staff and material to distant installations of water supply
- Health services, who have to control drinking water quality
- Security services for protection of installations of water supply
- Emergency services for mitigation of consequences

Figure 8.7. Break down of electric power supply has serious impact on water supply (© OÖ Wasser).

3. Preparedness for emergency situations in water supply

3.1. MEASURES FOR PROTECTIVE SECURITY AND PLANNING FOR CONSEQUENCE MANAGEMENT

The facts mentioned above about potential impacts and threats to water supply and real incidents should convince us that measures for protective security and planning for consequence management are necessary in the domestic water supply.

The first step must be a risk assessment, related to the special situation of the water supply system and to the relevant threats.

Based on the results of risk assessment it is necessary to create scenarios, pointing out consequences of relevant pressures on quantity and quality of drinking water. For this purpose the possible impacts, their interference and the vulnerability of facilities and elements of water supply have to be regarded under the supposition of a realistic worst case situation. The results of the risk assessment and the scenarios have to lead to concrete measures.

3.1.1. Protective measures

- Security of buildings and facilities – alert systems
- At least two "main pillars" of water abstraction, most suitable from different resources
- Redundancy in electric power supply and control-systems (i.e. emergency power system)
- Availability of disinfection systems for emergency cases

3.1.2. Preparations for recognition of impacts

- Training of members of staff to identify impacts in an early stage
- Clear definition of the way of notification of observances to the responsible persons

3.1.3. Preparation of response mechanisms

- Clear responsibilities for decision making – short ways
- Building of capacities of material and equipment
- Training of the staff for emergency situations
- Standard procedures for relevant and important impacts (only give guidance not to forget important things, no inflexible instruction)

- If emergency situations exceed a certain level, external support will get necessary. Arrangements for such situations should be made in advance
- Standard procedure for documentation of occurrences and reactions, as well for learning from experience as for purposes of later forensic activities
- Exercises, exercises ...

3.1.4. Information and communication

- In situations of crisis it is usual – but very dangerous for creditability – at first to say "nothing has happened" and then step by step to have to admit the full dimension of the emergency situation
- Information and communication have to be the job of specially trained persons
- Information and communication have to serve media as well as customers
- Information has to be given already at an early stage of the emergency situation and with full openness
- It is useful to prepare standardized information in advance, under stress there grows the danger of mistakes
- The better communication with media and customers is in "normal times", the more trust will be put in information under emergency conditions.

3.1.5. Support from outside the water work

- A crisis lab for "non target analysis", which allows to look weather something has happened, not really knowing what to look for
- Mobile purification plants – run by the Army or by NGOs
- Transport units for drinking water
- Machinery for packing water for transport and distribution
- Security services for times of crisis (police or army)

All preparations and precautionary measures have to be kept as simple as possible, otherwise they will not work under the stressing conditions of an emergency situation. And they also will not work in the case of emergency if they were not trained by exercises.

Of course exercises on this special topic are done well in the army and the emergency organisations. Both have also great experience from a great number of international deployments. But they take place more or less isolated, without involvement of water works. Emergency exercises in cooperation with water works are very rare. This is not only an Austrian problem, but also can be noticed at great international exercises.

One reason for it may be to avoid harassment of the public. But it would be very necessary to intensify this cooperation in exercises of all dimensions. The most important goal of exercises should be to increase awareness of enterprises of water supply for emergency situations, inform the public about risks and solutions concerning water supply and to improve the cooperation of the enterprises with emergency organizations.

3.2. LEGAL MEASURES AND GUIDELINES

The threats, the risk analysis is based on, have naturally different probabilities of occurrence – and in their effects different dimensions. Provisions of law addressing the providers of water supply systems cover the precaution against impairments of the water quality and quantity of water by "everyday" influences.

A comprehensive water protection policy including a lot of measures, among them protective areas for abstraction points of water supply systems proved successful, to maintain groundwater in a good status under "normal circumstances".

For threats with low probability and impacts of great dimensions no legal provisions exist for the precautionary measures. To fill this gap at least partly, the guideline "drinking water emergency supply" was elaborated in the framework of the Austrian Association of Gas- and Waterworks (ÖVGW) and decreed in the year 1988, revised in the year 2006. This guideline contains the entire spectrum of threats and the description of the measures which can be used as precaution and/or response. The guideline is intended to be implemented in the direct responsibility of the enterprises.

The guideline doubtlessly improved the preparedness of the water suppliers. Planning was accomplished, equipment purchased and cooperation agreed with emergency services. Altogether large water suppliers are usually better and more comprehensively prepared for crises, than smaller ones. The reasons for it are their financial power and the available human resources. Cooperation of water suppliers can prove helpful especially for the smaller ones to improve preparedness.

External support, in particular by the Federal Army, fire-brigades and the Red Cross have great relevance in this context as these organizations can offer important support to the water suppliers in extreme situations.

At this point I want to put emphasis on the point, that the Austrian Federal Army with AFDRU (Austrian Forces Disaster Relief Unit) and the Red Cross already in numerous cases gave aid in cases of natural disasters by autonomously operating water supply units at international missions. I would like to mention Armenia, Poland, Mozambique, Sri Lanka and Pakistan. So they have gathered special experiences in this sector, very helpful as well in the home country (Figures 8.8 and 8.9).

Figure 8.8. Austrian Forces Disaster Relief Unit (AFDRU), water purification and supply unit – deployed in Mozambique at the flood disaster (© ABC-AbwS).

Figure 8.9. Distribution of purified drinking water in Mozambique (© ABC-AbwS).

4. Lessons learnt from the flood disaster 2002 in Austria

The system of emergency water supply had to stand the test, when in 2002 great areas of Lower and Upper Austria were struck by a flood, which lasted several days in some regions. A great number of wells were contaminated by dirty surface water, sometimes polluted with heating oil from private houses. Pipes bridging rivers and ditches were cut off by the streaming flood and the electric power supply was interrupted.

Despite of these facts, a sufficient water supply could be held up by means of emergency water supply according to the guidelines mentioned above (Figure 8.10).

Keywords for it are

- Flying pipe connections with intact supply systems
- Mobile chlorination systems as disinfection for emergency cases, connected to the supply system to grant drinking water quality,
- Use of bottled water, produced by special equipment
- Drinking water distribution by fire brigades

Figure 8.10. Supply of people enclosed by water in their houses with commercial bottled water from a boat (© OÖ Wasser).

Figure 8.11. Disinfection systems for emergency cases – mobile chlorination (© OÖ Wasser).

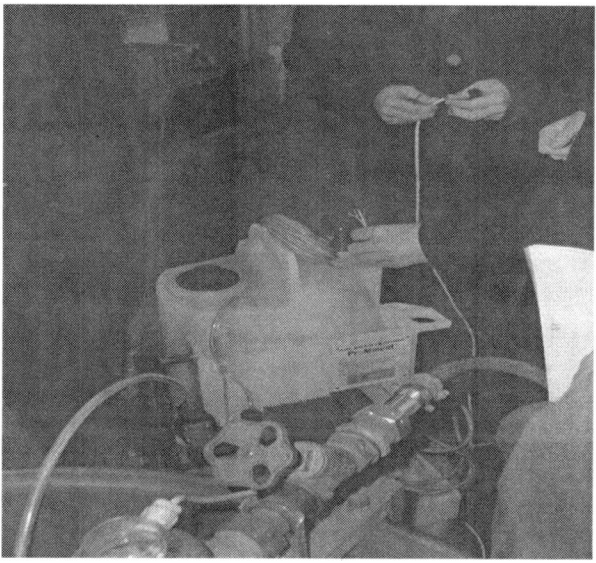

Figure 8.12. Disinfection systems for emergency cases – mobile chlorination (© OÖ Wasser).

- Employment of mobile purification plants run by the federal army (Figures. 8.11 and 8.12)

Water packages produced by special machines containing 1 l water in plastic bags, stabilized with silver compounds, proved more favourable for water distribution than buckets and cans of unknown hygienic condition, brought by people (Figures 8.13 and 8.14).

Figure 8.13. Production of water packages (© OÖ Wasser).

Figure 8.14. Production of water packages (© OÖ Wasser).

Of course there were lessons to be learnt from this flood disaster, but it got clear, that the level of preparedness made it possible to assure sufficient drinking water for the public in the indispensable extent. Of course it will also be the task of future to improve preparedness in the sector of emergency water supply (Figures 8.15 and 8.16).

Figure 8.15. Transport of the water packages in great boxes (© OÖ Wasser).

Figure 8.16. Distribution of water packages (© OÖ Wasser).

Of course there were a lot of lections to be learnt from this flood disaster, but it got clear, that the level of preparedness made it possible to assure sufficient drinking water for the public in the indispensable extent. Of course it will be also the task of future to improve preparedness in the sector of drinking water emergency supply.

5. Closing remarks

A workshop like "Threats to food and Water Cain Infrastructure" is the perfect platform for experts of different countries and of different areas of expertise for exchanging ideas and experiences. By this it offers a perfect technical contribution for preparedness for emergency situations.

Threats, risks, emergency situations and problems of security are issues, neither people, nor media nor politicians like much to talk of. But we have to be aware, that we are just now in the time before the next emergency situation. It is the duty of experts to put emphasis on the needs and not to get tired to address and underline them.

Let us act accordingly.

INTEGRAL MANAGEMENT OF WATER RESOURCES IN CROATIA: STEP TOWARDS WATER SECURITY AND SAFETY FOR ALL

ŽELJKO DADIĆ AND MAGDALENA UJEVIĆ

*Croatian National Institute of Public Health,
Rockefeller st. 7, Zagreb, Croatia
e-mail: zeljko.dadic@hzjz; magdalena.ujevic@hzjz.hr*

KSENIJA VITALE

*School of Public Health "A. Stampar", Medical School,
University of Zagreb, Rockefeller st. 4, 10000, Zagreb,
e-mail:kvitale@snz*

Abstract: The article investigates the number and the technical conditions of local water supply systems in Croatia. So far they have been poorly maintained, delivering water that is both microbiologically and chemically contaminated, to around 6% of Croatian population. Even when possible, users of local water supply systems prefer to stay outside bigger, communally controlled water supply systems, due to the pricing. Implementation of Water Safety Plans is recommended, as possible tool for improvement of the local water supply systems weather through merging with bigger systems, remaining as local units or preserving as an additional system in time of need. In addition these plans would help to classify the possible risks of each local water supply system and identify possible burden of disease as a consequence. That information could be point of arguing with local governments and communities. Better technical conditions of local water supply systems and sanitary acceptable water that is delivered are the starting point for the creation of alternative water sources network, essential for the crisis mitigation, by mapping preserved or those still in use.

Keywords: Croatia, drinking water, local water supply systems, network, water security and safety

1. Introduction

Demand for safe water and a sufficient quantity is becoming major environmental concern of our era. Well established relationship between clean water and good health, today, places water protection and management within the domain of primary health care. Croatia is considered water-rich country although resources are not spread evenly. Renewable sources, calculated as country's own resources, plus upstream flows, are around ten times more than in EU-15, with the considerably less withdrawal at the same time.[1,2] Population in Croatia is predominantly supplied with water through large, communal water supply systems (around 80%), that cover all urban centers. Small local supply systems are predominant in isolated rural areas, urban outskirts, or islands (around 20%), while the difference makes the population that uses private wells or cisterns.[3,4] Major health concerns are associated with all the forms of water supply systems that are not monitored by certified authorities and constantly sanitary controlled.

There is no exact or single definition of small local water supply systems, but there are several characteristics used for their determination; number of users, type of technology used and type of management. In Europe, small is defined by more than 50, but less than 5,000 users or more than ten but less than 1,000 m^3 per day withdrawal.[5] When less than 50 users, system is considered very small, in addition to that, wells and cisterns are also defined as very small systems.[6] As for management small systems are usually managed by local community or individuals without proper training and means.

In Croatia small local water supply systems (LWSS) were stimulated as quick and cheap solution for small and rural communities, without foreseen downside, that such a construction is usually outside general development plans, of low technical quality, without documentation and licenses and with unclear ownership.[7] Adverse health potential of local supply seems underestimated, although they are the most vulnerable systems for microbiological or chemical pollution, both from natural and/or anthropogenic origin. Microbiological contamination is the most common due to the shallow surface accumulations, used for withdrawal, that depend on precipitations. Along with that another specific problem arises in coastal and island areas, and that is water scarcity, due to the increased tourist demand and climatic characteristics. Although number of waterborne epidemics, as an indicator of microbiological pollution, is relatively small, 26, affecting 1,734 people, in period from 1992 to 2006.[8] However, some authors[9,10] argue that number is larger due to underreporting and cases that never reach health care system. Most common outcomes were enterocolitis, disenteria, gastroenterocolitis, gastroenteritis, hepatitis A. Out of 26 epidemics, 24 were in the LWSS.[8] Even bigger problem could be long-term, low dose exposure to the chemical pollutants but detailed local longitudinal studies are missing.

The aim of this article/research is to determine the number and general conditions of small water supply systems in Croatia, along with the number of people that they serve, in order to create best possible mode of sanitation or potential of merging with big and sanitary controlled systems. Also, resolving the problems connected to the small water supply systems is one of the EU criteria.

2. Materials and methods

We listed all small local water supply systems in Croatia through the data base of national and county institutes of public health, local sanitary inspection, local governments and field visits where needed, that supply water for more than 50 users. This cut off point is adopted according to the currently used legislation[11] that defines as public supply all systems that serve more than 50 users or withdraw more than 10 m^3 per day.

Water samples are taken according to the standard procedure,[12] along with questionnaire covering technical and managerial details (methods of sanitation, number of employees, education of employees, number of people served etc.). Estimation of delivered water has been expressed as total of delivered water at the day of sampling, and it should be taken with caution. Water samples were tested for physical, chemical and microbiological agents according to the current legislative,[11] and standard methods.[13] Analyses were performed at National Institute for Public Health in Zagreb and County Public Health Institutes in Osijek, Pula, Rijeka, Dubrovnik, Zlatar and Varaždin.

3. Results

The results are presented in Table 9.1 showing that, 443 located, local water supply systems are serving 260,000 people which makes around 6% of population of Croatia. That leaves us with even higher number of people with individual systems of supply such as wells or cisterns (14%).

Local water supply systems in Croatia are mostly in Black Sea catchment, while the rest is part of the Adriatic sea catchment. Most of the local water supply systems are in Krapinsko–Zagorska (112) and Varaždinska (80) county.

TABLE 9.1. Local water supply systems in Croatia.

County	Number of local water supply systems	Number of springs	Estimation of delivered water (m³/day)	Number of people
Black Sea catchment				
Zagrebačka	54	92	2,583	12,251
City of Zagreb	25	36	1,670	10,400
Krapinsko–Zagorska	112	132	5,265	39,060
Sisačko–Moslavačka	12	19	200	1,330
Karlovačka	22	29	466	2,515
Varaždinska without Ivanac	80	99	3,372	18,440
Međimurska	2	2	240	1,461
Koprivničko–Križevačka	8	8	2,180	2,693
Bjelovarsko–Bilogorska	8	12	1,190	13,350
Virovitičko–Podravska	12	56	1,805	10,750
Požeško–Slavonska	6	6	90	983
Brodsko–Posavska	5	8	1,090	7,453
Osječko–Baranjska	29	33	4,970	44,469
Vukovarsko–Srijemska	28	26	5,861	46,886
Adriatic Sea catchment				
Istarska	6	6	70	440
Primorsko–Goranska	19	49	17,571	41,374
Ličko–Senjska	6	6	622	2,936
Zadarska without Biograd and Gračac	2	6	400	2,750
Šibensko–Kninska	4	4	1,150	1,570
Splitsko–Dalmatinska	1	1		3,000
Dubrovačko–Neretvanska without Blato	2	2	90	725
Total	443	632	50,885	264,836

All of them are managed by users who also financed the construction. Water is withdrawn from springs not evidenced in system of state or local water management hence without licences. Local water supply systems are not continuously sanitary monitored. Control is performed upon request and the estimation of the users, as well as timing and type of sanitation (chlorination). In Table 9.2, results of physical, chemical and microbiological analysis are presented showing serious problem of contamination.

When divided by tested agents, for physical and chemical agents 109 (20.2%) samples proved unacceptable and for microbiological agents 345

TABLE 9.2. Results of sample analysis.

	Number	Percent
Sanitary acceptable samples	135	25.1
Sanitary unacceptable samples	403	74.1
Total	538	100

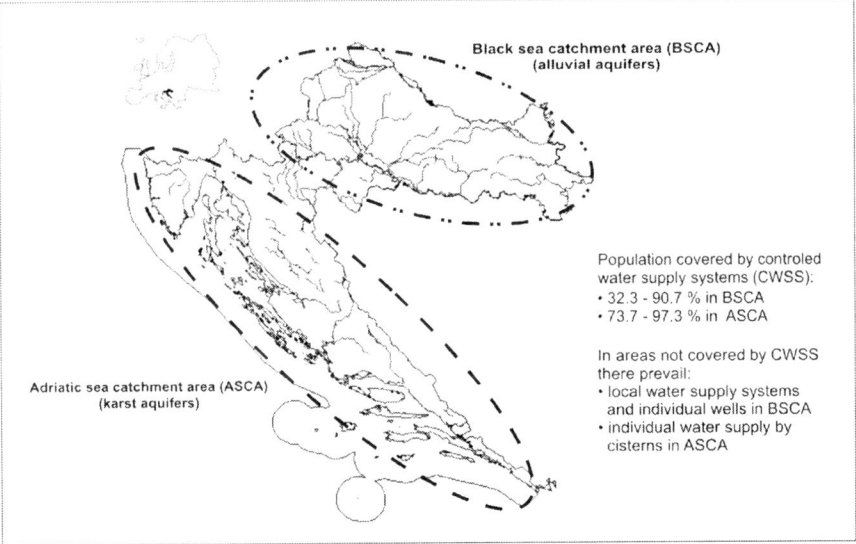

Figure 9.1. Water supply in Croatia according to catchment areas and corresponding counties.

(64.1%) samples proved unacceptable. Some samples were unacceptable for both chemical and microbiological agents, which resulted in the difference regarding the total number of sanitary unacceptable samples. Also we analyzed data by regions mainland, rich in water resources coming from alluvial aquifers (Panonic basin), while coast is supplied from karst aquifers. In central part of Croatia in Krapinsko–Zagorska county all samples were unacceptable due to the microbiological contamination. Similar situation is in Varazdinska county, with only 12% of unacceptable samples due to the physical and chemical contamination. In eastern part of Croatia, most of the contamination comes from physical and chemical agents i.e. turbidity, manganese, iron and arsenic leading to 96.1% in Osjecko–Baranjska county to up to 100% of unacceptable samples in Vukovarsko–Srijemska county. Only smaller part of unacceptable samples is due to the microbiological contamination. There are fewer samples from karst zone due to the fact that in those counties most of the users are connected to the big and monitored water supply systems (Figure 9.1). In this region microbiological agents are responsible for the most of the contamination, biggest percentage being in Primorsko–Goranska county (94.5%). In all counties microbiological contamination counted for 50% up to almost 100% of unacceptable samples which indicates strong need for disinfection and constant monitoring. As potential sources of contamination we identified drainage from unsanitary landfills and agricultural land and precipitations.

Despite the awareness of potential health hazard, it seems that most of the communities prefer current state of the water supply because it is almost free. According to the existing legislative[14] waters are "general goods" and if used without special technology for household needs (irrigation excluded) it is free. This article is loosely interpreted, leaving space for inertia of local government. On the other hand, in ever changing social conditions in Croatia, this could be hidden help for many of low income households.

4. Discussion

At present the overall annual infrastructure construction growth in Croatia is around 1%, with expected growth of up to 90% of coverage. Due to the geographical specificity, such as high mountains and distant islands with scattered households, communal water supply system is not economically sustainable everywhere.[15] Therefore, portion of the population will always remain outside the monitored system with various degree of risk. However, where possible, measures for sanitation and regulation of local water supply systems should be taken seriously. Our results correspond with the previous study of local water supply systems[7] identifying microbiological and chemical contamination. Our study identified smaller number of water supply systems by 79 units, which indicate that some are shut down or preserved, but on the other hand, number of the users has grown for almost 120,000 people. This means that substantial number of people is constantly connecting to existing LWSS with tendency to overload accessible capacity of withdrawal and distribution. Also it indicates problems in systems of the construction control and control and management of water resources. Even when possible people do not connect to controlled water supply systems (CWSS) most probably due to the high water prices.

Traditionally in Croatia water is managed on the "command and control" approach with system of laws and standards that prescribe the way of behavior in "top down "manner. This approach is suitable for big water projects, but is seems that it failed in management of LWSS. Since local authorities should be main forces in the proceses of development of local environment, health and social determinants different approach should be implemmented, offering "middle road" (both taxes and subsidies) acceptable for all.

The water prices in Croatia are at the present social category, and they are not reflecting real costs of delivery. They vary from 0.6 to 10 euro for cubic metre, with average between 0.5 to 2.1 euro, depending on the calculation of the local authority (the highest prices of 10 euro are charged to non-residents, summer or second house owners, on some islands). The projections are, that, it will increase until 2015 to around 3 euro for cubic metre and should be equal for all users. However, according to the European directive price of the water should not exceed 4% of average salary and that Croatian waterworks company recommends this as the target price.[16] How this will be applied and how it will affect poorer communities, still remains to be seen. Experience has shown that progress is possible only if all stakeholders are motivated and mobilized in the project, so this is another challenge in LWSS managing. It seems that water supply is an issue where Croatia is facing a challenge of combining the economic and social responsibility. Problem of tension between the two approaches is no novelty, but water supply is a broader issue than just an economic category. As mentioned before, water supply is crucial for human health and it directly influences the health indicators, and consequently the health care cost. Health care is free and by law available for all the citizens, but some degree of individual responsibility has to be implemented. Both individuals and government have to take responsibility for health and health related actions, and solving of LWSS is one of such a projects. Berlin Declaration officially "Declaration on the occasion of the 50th anniversary of the signature of the treaties of Rome"[17] rests upon the priority on public health. The Declaration is important document because it foresees the – bigger than at the moment – differences in health indicators between the groups of lower and higher socio-economic status, and stresses the need to combat the disease and poverty by socially responsible economy.

Importance of water supply is reflected in "International Decade for Action: Water for life 2005–2015" launched by UN, with goal to implement innovative approach towards improving water sanitation. It has been proven that interventions such as point-of-use disinfection and educational programs reduce disease prevalence.[18] One of the approaches could be Small Community Supply Management Network (SCSMN) program; launched by WHO in 2005 that establishes principles and policy for resolving problems of small water supply systems.[19] Among key principles of the SCSMN is that the underlying ethical principles should ensure comparable levels of safety for small and large supplies. Framework for safe drinking water outlined by WHO,[20] comprises five components including (1) health goals, (2) system estimation, (3) monitoring, (4) management plan, (5) independent monitoring. Health goals, based on the WHO Guidelines on Drinking water quality,[20] are set by each country depending on individual situation and potentials. Components 2, 3 and 4 actually form a Water Safety Plan – WSP,[20] an

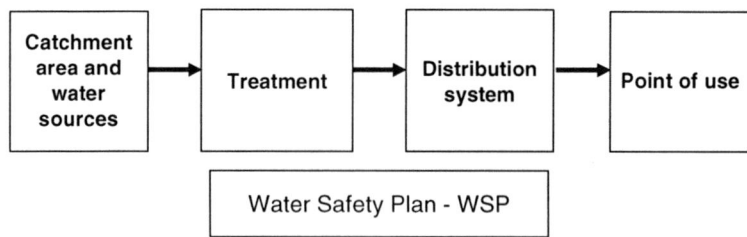

Figure 9.2. Schematic presentation of the "catchment to consumer" approach to risk management of the safety of drinking water (Adapted from Medema G J et al. 2003[21].)

innovative tool for assessing and managing potential risks. WSP encompasses all elements of production of safe drinking water from the catchment to the consumer (Figure 9.2). Each stage of the process is described in detail, forming written document that is understandable approach to the risk assessment and management. Possible risks are identified from the catchment area to the point of use, they are defined, prioritized, and in the end control mechanisms are offered.

Along with the WHO efforts to find out best tools for improvement of LWSS, EU legislative is changing in order to support those efforts. Accordingly, EU Directive on drinking water has been revised, and one of the new elements is application of water safety plans for LWSS.[22] Such plans could be of great importance in managing LWSS in Croatia, following the proposed steps, each LWSS could identify its priorities, possible risks and decide whether it could merge with CWSS, remain as single unit or if it could be shut down and preserved.

Infectious diseases has become a security issue, ranking water as possible mean of transmission, as a highly risk media. In such a scenarios big water supply systems could be primary target. Creation of the resource networks for the specific region (county or even state level) could be of enormous importance in the situation of the crisis, both natural or man made. Therefore locating of all the water resources (including cisterns, wells and LWSS) in the country, their state, level of needed reconstruction and capacity should be one of the goals in the preparedness plans. Network of LWSS could serve as temporary water supply system in time of crisis. It is well known that first response in the process of the mitigation is crucial for the survival and following epidemiological development (incidence of infectious diseases). Water resources that could be quickly mobilized such as previously preserved LWSS should be one of the tools in crisis respond and its network creation one of the goals of local governments.

Systems that are not fully organized in terms of not knowing its own advantages and downsides are always most vulnerable point in economic, social or health crisis.

Bibliography

1. National strategic plan of environment protection (in Croatian). National Gazette 2002; No. 46.
2. Geres D, Mijatovic I, Dadić Z, Lovric E, Sustainable water use in Croatia (in Croatian). In: Lovric E, editor. Proceedings of the 5th symposium on water and public water supply; 2001 Oct 3–6; Tučepi, Croatia. Zagreb: Croatian Institute of Public Health; 2001. pp. 107–126.
3. Croatian Water Management Strategy. Croatian Waters, 2007, draft.
4. Gereš D, Dadić Ž, Lovrić E, Water use for public water supply in Croatia (in Croatian). In: Lovrić E, editor. Proceedings of the 9th symposium on water and public water supply; 2005 Oct 3–6; Osijek, Croatia. Zagreb: Croatian Institute of Public Health; 2005. pp. 5–23.
5. Council Directive 98/83/EC on the quality of water intended for human consumption. Official Journal, L 330, 05.12., 32, European Commission, 1998.
6. Hulsmann A, Small systems large problems: A European inventory of small water systems and associated problems. Web-based European Knowledge Network on Water. Web-based European Knowledge Network on Water/ENDWARE. 2005. Available at: http://www. weknowwaternetwork.com/uploads/booklets/05_small_water_systems_ver_june2005.pdf
7. Afric I. The role of water sanitary inspection in the health care quality: Example from the island of Krk. (in Croatian), Master thesis, Medical School, University of Zagreb, 2005.
8. Dadić Ž, Lovrić E, Ujević M, Ambrenac J, Gereš D, Small drinkinking water supply systems – public health risk (in Croatian). In: Dadić Ž, editor. Proceedings of the 11th symposium on water and public water supply; 2007 Oct 3–6; Bol, Croatia. Zagreb: Croatian Institute of Public Health; 2007. pp. 5–17.
9. Smoljanović M, Waterborne epidemics– main thoughts (in Croatian). In: Dadi Ž, editor. Proceedings of the 11th symposium on water and public water suply; 2007 Oct 3–6; Bol, Croatia. Zagreb: Croatian Institute of Public Health; 2007. pp. 31–53.
10. Cipriš R, Kunović M, Ištok J, Local water supply systems in Krapinsko-zagorska county (in Croatian). In: Dadić Ž, editor. Proceedings of the 11th symposium on water and public water suply; 2007 Oct 3–6; Bol, Croatia. Zagreb: Croatian Institute of Public Health; 2007. pp. 17–23.
11. Pravilnik o zdravstvenoj ispravnosti vode za piće, Narodne novine br. 47/2008. (Drinking water regulation. National Gazette 2008; No. 47).
12. HRN ISO 5667-5:2000, Kakvoća vode -- Uzorkovanje -- 5. dio: Smjernice za uzorkovanje pitke vode i vode za pripremu hrane i napitaka (ISO 5667-5:1991) Water quality – Sampling – Part 5: Guidance on the sampling of drinking water and water used for food and beverage processing (ISO 5667-5:1991).
13. Standard methods for the examination of water and wastewater, 21th Edition, American Public Health Association, Washington, 2005.
14. Zakon o vodama, Narodne novine br. 107/1995 i Izmjene i dopune Zakona o vodama, Narodne novine br. 150/05. (Water act, Official Gazette No. 107/1995; Water act amendments, Official Gazette No. 150/05).
15. Dadić Z, Reference book on basic water quality in Croatia (in Croatian), Croatian National Institute for Public Health, Zagreb, Croatia, 2001.
16. Zagreb communal gazette. Zagrebacki komunalni vjesnik (In Croatian) from 26th of March 2008 number 370, p. 4.

17. Berlin Declaration officially "Declaration on the occasion of the fiftieth anniversary of the signature of the treaties of Rome" Available at www.eu2007.de/eu/About_the_EU/
18. Hughes JM, Koplan JP. Saving lives through global safe water. EID 2005; 11:1636–1637.
19. Report of the First Meeting of the Small Community Water Supply Management Network, Reykjavik, Iceland 24–26 January 2005. WHO 2005. Available at: http://www.who.int/water_sanitation_health/dwq/scwsm_network/en/index.html (Access 01.09.2008)
20. WHO. Water safety plans. Managing drinking-water quality from catchment to consumer. World Health Organisation, Geneva, 2005.
21. WHO. Guidelines for Drinking-water Quality. Third Edition. World Health Organization, Geneva, 2004.
22. Medema G J , Payment P, Dufour A, Robertson W, Waite M, Hunter P, Kirby R, Andersson Y, Safe drinking-water: an ongoing challenge. In: Dufour A, Snozzi M, Koster W, Bartram J, Ronchi E, Fewtrell L, editors. Safer drinking-water: improving the assessment of microbial safety. IWA Publishing, London, 2003. pp 11–45.
23. WHO Europe, 2007. Support for the Development of a Framework for the Implementation of Water Safety Plans in the European Union. Version 4 October 2007.

TERRORISM AND THE FOOD CHAIN

EVELYNE MADELAINE MAILLOT

Veterinary Public Health, General Counsel for Agriculture, Food and Rural Areas, Ministry of Agriculture and Fisheries, France
e-mail: evelyne.maillot@agriculture.gouv.fr

Abstract: A study concerning deliberate threats against the food chain has been carried out in France using various methods: analysis of feasibility and impact on public health; collection of data on previous biological and chemical outbreaks, whether malevolent or accidental; and the vulnerability of the food chain. Twenty-four biological agents have been classified in three categories of increasing danger. The same approach has been used for 71 chemical agents. It should be noted that the terrorist risk to the food chain, especially the biological risk, must be seen in context. In the case of chemical agents, it must be borne in mind that they are not much affected by the temperatures of foodstuff treatments and that some are readily available on the market or on the internet.

In food factories, the Hazard Analysis Control Critical Points method is aimed at controlling sanitary risks and preventing accidental outbreaks. This kind of method can also be developed to prevent criminal activities. Each food factory can implement its own specific programme for the prevention and management of threats, restricting access to equipment, premises and products, and monitoring people's movements. A practical guide book to help French food producers to prepare their own prevention plan is now available.

Keywords: Food chain, terrorism

1. Introduction

Within the agricultural sector, terrorism may target foodstuffs, water, zoonoses, animal health and crops. Public health, the economy and food supplies may be affected, and this can give rise to panic and disruption of services.

V. Koukouliou et al. (eds.), Threats to Food and Water Chain Infrastructure,
DOI 10.1007/978-90-481-3546-2_10, © Springer Science+Business Media B.V. 2010

In 2003 a study was begun in France on the risks to public health via the food chain. It consisted of three levels: risk assessment, prevention, and preparation for dealing with an attack.

The author focused on the risk of contamination of the food chain and on producing a guide for use in the food industry.

2. Assessment of risk to the food chain

The risk of a potential terrorist attack can be assessed by studying the characteristics of dangerous substances which may be used or by examining past cases of accidental or criminal food poisoning, with reference to the vulnerability of the food chain.

3. Study and classification of dangerous substances

A study carried out in France in 2003–2004 looked at biological and chemical agents which pose a danger to public health and for which a risk rating has been established on the basis of analysis criteria in numerical form.

The assessment was made by a group of experts from the Ministry of Agriculture, the Institut Pasteur in Paris, the army, and French public health agencies involved in monitoring and evaluating the safety of food, medicines and human health.

The group established an initial list of agents to be assessed on the basis of lists from studies carried out abroad, on similar but not identical subjects, by the *Australia Group* (which works to prevent the proliferation of biological and chemical weapons through exports), *the European Union* and the *Center for Disease Control* (CDC) in Atlanta, USA.

As regards biological agents (viruses, parasites, bacteria et toxins), all agents known to have been the source of unintentional incidents were included, with the most pathogenic strain when there were several. The overall scenario is one of home-made production, excluding state terrorism. Exotic or genetically modified agents or strains were excluded owing to lack of data.

As for chemical agents (natural or synthetic, metallic, toxic, hallucinogenic and hypnotic, corrosive, pesticidal), a pre-selection was made among those which cause acute symptoms (rapid effects) and which are transmitted mainly via foods.

Three groups of criteria were considered: impact on public health, feasibility, and moderating factors, with quantitative date (mortality rate, hospitalization rate) or qualitative data with scores of 0 to 2 (e.g., for ease of preparation: easy, medium, difficult).

Finally the agents were classified in three groups according to the degree of danger. Biological and chemical agents were treated separately.

Of the 24 biological agents studied, using 15 criteria, two were put in Class 1 (*anthrax* and *botulinum toxin*), eight in Class 2, and six in Class 3. The other eight agents were considered to be of negligible risk. Those agents which are the source of the most serious and most common types of collective food poisoning are in Class 3. Finally, another reassuring result is the fact that no biological agent has both a strong impact on public health and great ease of preparation and use.

Of the 71 chemical agents studied, using seven criteria, 13 were put in Class 1. The scores of the chemical agents fall within a fairly narrow range, with little difference between the three classes.

The detailed report on this work is classified.

4. Past cases of accidental or criminal food poisoning

There is no record of any terrorist attack on the food chain. In order to estimate the potential impact of such an action, we can look at past cases of criminal food poisoning and accidental cases.

Table 10.1 shows the main known cases of criminal contamination of food throughout the world. Table 10.2 shows the most serious cases (in terms of casualties) of accidental contamination of food.

Fortunately, the numbers of casualties (illness or deaths) reported in Table 10.1 (main known cases of criminal contamination of food), are not very high, and much lower than in major accidental poisonings.

In Table 10.2, the very high numbers of cases of salmonella and hepatitis A are due to the multiplication of these biological agents in a large quantity of food at a favourable temperature and over a long period. It is not easy to combine these three conditions deliberately, and it is therefore unlikely that terrorists could achieve results on such a large scale.

Comparison of the two tables confirms the results of the study on dangerous substances: the risk of bioterrorism against the food chain seems less important than it looks like on first instance. In particular, we should consider the complexities involved in acquiring, producing and conserving biological matter which is naturally fragile, and in using it successfully in food.

Chemical agents are more stable, particularly at food treatment temperatures, and some are readily available in shops or on the Internet.

TABLE 10.1. Cases of criminal contamination of food products.

Date	Products involved	Type of contamination or contaminant	Consequences	Aim Observations
1972		40 kg of Salmonella Typhimurium culture	None: preventive action taken by the police	Two members of the "Order of the Rising Sun" arrested in Japan in possession of the products
1977	Citrus fruits from Israel	Mercury, probably injected by syringe	≃12 people contaminated Severe drop in exports from Israel	Damage to the Israeli economy
1980s	Drinks and various foodstuffs in Iraq	Thallium	Several dissidents poisoned	Elimination of political opponents
Sept–Oct 1984	Salads (ten restaurants in the same chain) in Oregon (USA)	Liquid Salmonella typhimurium culture	750 people poisoned and 45 hospitalized	Religious sect attempting to influence a local election
1989	Chilean grapes imported into the USA	Cyanide	No-one contaminated. Several countries suspended imports of Chilean fruit	Damage to the Chilean economy
1991	Perrier water in France	Traces of benzene in several bottles	Products recalled 35% fall in turnover	Aim and culprits unknown. Contamination may or may not have been accidental
1992	Water tanks at a Turkish military camp in Istanbul	Lethal concentrations of potassium cyanide	Officially, no military personnel contaminated	Poisoning of garrison by the PKK
1995	Champagne (Russian military camp in Tajikistan)	Cyanide	≃10 Russian military personnel killed	Afghan revenge? Withdrawal of Russian army from the country?
Oct 1996	Cakes (staff room for laboratory personnel at an American medical centre)	Shigella Dysenteriae type 2 (from lab cultures)	12 people contaminated, of whom four hospitalized (from a team totalling 45)	Revenge of an employee

Date	Food	Poison	Consequences/Response	Notes
1996	Various foodstuffs in various food groups in the FRG	Snake venom (cobra and viper)	More stringent checks. Crisis cells	Extortion of 400 m DM in diamonds by a mysterious commando named Tamara S. Case never solved
2002	Breakfasts at a fast-food establishment in China	Rat poison	40 deaths including some children, 200 people hospitalized	Business competition
Early 2003	Soya milk in eight primary schools in northern China	Not announced	3 child deaths, more than 3,000 poisoned	"Criminal poisoning" with no further details; or fraud?
July 2004	Drinks, chocolates and cheeses from six industrial groups in France	Not determined	Crisis cells for each group. Intergroup cell with the police	Attempt at extortion by the mysterious "AZF group" (case known as "AZF 2")

N.B.: Non-exhaustive list limited to cases reported in the press.

TABLE 10.2. Most serious cases of accidental food poisoning.

Biological agents

In the USA in 1985, about 170,000 people were infected with *Salmonella typhimurium* from milk which had been contaminated in the factory after pasteurization (the source was the interior of the pipes).

In China in 1991, about 300,000 people contracted hepatitis A from infected clams.

In the USA in 1994, about 224,000 people were infected with *Salmonella enteritidis* from ice cream.

In Japan in 1996, about 8,000 children fell ill after eating radishes at school which were contamined with *E. coli* 0157: H7, resulting in several deaths.

In Denmark in 1998, about 10,000 cases of salmonellosis were linked with consumption of industrial mayonnaise.

Chemical agents

In Japan in the 1960s, 200 people suffered mercury poisoning after eating fish from polluted waters.

In Iraq in 1971, more than 6,500 people were hospitalized with nervous system symptoms and 459 died after eating bread made with mercury-contaminated wheat.

In Spain in 1981, about 20,000 people were poisoned and 800 died as a result of cooking oil which had been polluted with a chemical agent.

In the USA in 1985, 1,400 people fell ill after eating melons which had been grown in fields treated with the pesticide aldicarb.

5. Vulnerability of the food chain

The vulnerability of the food chain is linked with the nature of the food and with the target.

In food, the concentration of infectant must be sufficient to produce symptoms. In particular, the noxious product must be well distributed in the target food, so the consistency of the food (homogeneity, fluidity) and its composition (fats, proteins, sugars …) have a role to play. In addition, the number of casualties will depend on the amount of contaminated food and thus on the size of the batch.

Stability in the foodstuff is also important, together with the temperature, the humidity of the food, the pH, and also the length of time before consumption.

Certain consumers and/or establishments may be particular targets in view of their symbolic value (e.g. young people, links with a community, airline, flagship brand), or because production is high (large numbers of consumers), or because there is easy access to the food at the establishment or in pre- or post-production (suppliers, transport, distribution, storage) and people's movements are not monitored.

The American CARVER method assesses the vulnerability of an enterprise and prioritizes the risks and consequences resulting from a malicious act, starting from the most vulnerable, attractive and accessible points for a potential attack. The method involves breaking the process down into units which may represent a vulnerable point and then attempts to assign a value to each one using the following criteria:

Criticality: estimation of the consequences of an attack in terms of public health (number of potential deaths) and economic impact

Accessibility: estimation of the ease of access to the target and the ease of leaving it after an attack

Recuperability: estimation of the potential for recovery after an attack

Vulnerability: estimation of the ease of carrying out the attack

Effect: estimation of the direct losses resulting from an attack, measured by loss of production

Recognizability: ease of identification of the target

With this method, a "critical point" in an enterprise involves a combination of four factors which bring the food and the target into play: large contaminable batch, short shelf-life of the food, uniform dilution of contaminant in the food, easy access to the food.

6. Prevention of terrorist risk against the food chain general health and safety measures

General health and safety measures are designed to prevent accidental contamination by avoiding the unintentional introduction, persistence, multiplication or diffusion of biological or chemical agents. This requires health and safety controls at all levels of production, from the farm to the fork.

General health and safety, which is mandatory, is in many ways a basic protection against malicious, criminal and terrorist acts, even though it is designed to avert a different type of risk – the accidental risk. Thus, maintaining the cold chain is intended to prevent the multiplication of biological agents, regardless of whether they find their way into the product accidentally or intentionally. The principle of product traceability enables us to monitor sources.

The European regulations (Food Law 2002 and Hygiene Regulations) filter down into every food establishment as a health and safety management plan. The operator establishes the plan himself on the basis of his risk assessment using the HACCP method – Hazard Analysis Control Critical Points. Good practice guidelines for professionals clarify the measures that should be taken.

There are also monitoring and inspection systems for the use of professionals and the official services.

7. Specific prevention measures

Measures designed to protect against terrorist, criminal or malicious acts using NRBC agents in food enterprises are voluntary (not mandatory).

Each food establishment is invited to draw up its own security management plan based on an internal assessment of vulnerability.

Several methods may be used for this assessment. The VACCP method (Vulnerability Analysis Control Critical Points) is similar to the HACCP which is used to draw up the health and safety plan. With the more sophisticated CARVER/SHOCK method, it is also possible to prioritize the risks and consequences of a potential attack.

In July 2007 the French Ministry of Agriculture and Fisheries produced a guide to this process and distributed it to professional food organizations: "Guide to recommendations on protecting the food chain against the risk of malicious, criminal or terrorist acts".

The guide is intended to:

- Prevent intrusion by third parties, malicious actions by staff or partners, the introduction of dangerous products or products which have been tampered with, and the theft or use of toxic products held legitimately

- Maintain and monitor the integrity of raw materials, intermediate products, finished products, packaging and processing, during all stages of preparation, transport and storage of the foodstuffs

- Enable users to deal rapidly with any alert or anomaly detected

It makes numerous general recommendations relevant to all establishments on the physical protection of access points, monitoring the movements of individuals, vehicles and products, stock control, information security, and special recommendations for the collection, import, transport, catering and distribution sectors. It outlines the assessment methods mentioned above and the preparation of a business security management plan. This plan must also be adaptable to meet different levels of threats.

Moreover, in 2006 the food sector was designated as an area of vital activity (for the country) in France, as it was in the European Union. This meant that several major food operators could be designated "operators of vital importance"; in these cases, the preparation and implementation of protection measures will be obligatory.

8. Preparing to deal with an attack on the food chain

In France, the "guide to dealing with an alert in the food sector when a product or batch has been identified" has been covering accidental incidents for some years. This guide is also relevant to the terrorist risk. It deals with the assessment of the situation, assessment of the danger, withdrawal and recall of products, stepping up controls even outside the range of the alert, communication and coordination of efforts.

In addition, the public health agencies have produced special leaflets on diagnosis and treatment in the event of a terrorist or malicious act using one of the main dangerous substances. A study has also been carried out on the requirements of certain specific treatments such as antidotes.

9. Conclusion

There is no record of any terrorist attack on the food chain, and previous studies suggest that the risk is more moderate than usually thought. Nevertheless, this subject must be taken seriously and preparatory work must be done.

In France, the ministry in charge of agriculture organized a risk assessment for the food chain and then produced a guide for food professionals to help them organize their own prevention. The protection of water tanks was dealt with similarly. Work on the risks to crops is underway.

In addition to the risk of deliberate contamination of foodstuffs in pursuit of terrorist or malicious intentions against human or animal health or against cultures, there is another type of risk to food – deliberate contamination for economic or fraudulent reasons, such as cooking oils adulterated with engine oil, or melamin used as a substitute for protein – and this should be analyzed and managed differently.

WATER FOR VIENNA

ASTRID ROMPOLT

*Stadt Wien – Wasserwerke, Grabnergasse 4-6, 1060
Vienna, Austria
e-mail: astrid.rompolt@wien.gv.at*

Abstract: Whereas most of the world's large cities have to cover their water
requirements from rivers or ground water reservoirs, Vienna is able to get its
drinking water from Alpine springs: pure Alpine Springwater in its natural
state is conducted from its sources in the mountains to the city. This water
is almost miraculously pure, especially considering the environmental pollution, which is so widespread today.

Map of Watersupply: http://www.wien.gv.at/english/environment/watersupply/
images/hql-plan-eng.gif

The foundation of Vienna's modern water supply system was laid long
ago with the construction of the First Vienna Alpine Springwater Main,
which has its source in the mountain range of Rax-Schneeberg-Schneealpe,
and the Second Vienna Alpine Springwater Main, which has its source in the
Hochschwab area. These water mains rightfully enjoy international renown
and no doubt figure among the municipality's major achievements.

Keywords: Vienna waterworks, water distribution, water supply, vienna water quality

1. Water distribution

Balancing natural resources and varying water demand in the city is one of
the central tasks of the Vienna Waterworks. For this purpose, it operates 30
reservoirs (28 of these situated in Vienna) with a total capacity of 1.6 million
cubic metres, which roughly corresponds to a 4-day consumption volume.

Map http://www.wien.gv.at/english/environment/watersupply/images/
vienna.gif

The map shows the main pipeline system of the Vienna Waterworks and
the pressures zones of Vienna's water supply.

The pipeline network is composed of several pressure zones, which are
a consequence of the different altitude levels of the areas supplied. The First

V. Koukouliou et al. (eds.), Threats to Food and Water Chain Infrastructure,
DOI 10.1007/978-90-481-3546-2_11, © Springer Science+Business Media B.V. 2010

Mountain Spring Pipeline supplies the topographically lower areas of Vienna (blue, violet), while the Second Mountain Spring Pipeline handles the areas situated at higher altitudes (red, green, orange, brown and yellow) in the western part of Vienna. Those areas where natural water pressure alone is insufficient (yellow) are supplied by means of pumping stations. The water pressure in all pressure zones is constant at 3–5 bar.

The water reservoirs store the mountain springwater fed into the system and keep it fresh through continuous flow-through. The oldest reservoirs date from the nineteenth century (e.g., the Rosenhügel Reservoir was completed in 1873), while the most recent one – the Jubiläumswarte Reservoir – was officially inaugurated in autumn 2006.

2. A brief history of the Vienna water supply system

2.1. THE BEGINNINGS

Already the Romans collected springwater in the area which is Perchtoldsdorf and Gumpoldskirchen south of Vienna today and conducted it to Vindobona. Medieval times witnessed a setback: Up to the sixteenth century the population drew its water exclusively from local house wells. The first documented water main was the "Court Main of Siebenbrunn", which was built around the year 1553 and served to supply water to Vienna's Hofburg and some of the buildings in the city center. The oldest municipal water main, the Hernals water main, was built in 1565. It delivered water from today's 17th district to a well house on Hoher Markt.

2.2. EXTENSION OF THE WATER SUPPLY SYSTEM

Even though a few smaller mains were built, these measures failed to relieve the water shortage in Vienna. In the eighteenth century, the "water man" or the "water woman" traveling the streets and offering water from a barrel became customary figures in Vienna's city life. The city had some 10,000 house wells, which were frequently contaminated. This situation started to change only when the Albertinian Main (1804) and the Emperor – Ferdinand Main (1841/46) were built, securing a more adequate water supply to the city. However, with the incorporation of the suburbs starting in 1850, Vienna developed into a large city and definitively outgrew its old water supply system. In this situation, the authorities took a far-reaching decision which would solve all problems: they decided to build the First Vienna Alpine Springwater Main.

2.3. THE FIRST VIENNA ALPINE SPRINGWATER MAIN

In 1864 the Vienna city council resolved to build the First Vienna Alpine Springwater Main, which still covers about 40% of Vienna's water requirements. This main was to secure adequate water supply also to the suburbs and help to improve the quality of the water to a point where it no longer represented a health hazard for the population.

2.3.1. Inauguration in 1873

Following a construction period of only 3 years, Emperor Francis Joseph I on October 24, 1873, inaugurated the First Vienna Alpine Springwater Main with the ceremonial opening of the high-jet fountain on Vienna's Schwarzenbergplatz. The main, which was built at a cost of 16 million gulden, has nowadays a length of 150 km and became a symbol of the liberation from water shortage and the threat of epidemics. For city dwellers, the former house wells were replaced by the so-called Bassena, supply taps in the corridors of houses. Already in 1888, more than 90% of the inhabited buildings within Vienna's city precincts were connected to the new main.

2.4. SECOND VIENNA ALPINE SPRINGWATER MAIN

Notwithstanding the expansion of the First Vienna Alpine Springwater Main, the city was again faced with supply bottlenecks in the wake of the municipalization of the suburbs (1890/92). Following extensive preparatory work, the city laid the foundation for the Second Vienna Alpine Springwater Main in 1900. The main was built under Mayor Karl Lueger.

Ten thousand workers were employed in building this main with a total length of 180 km, which serves to transport water from springs in the Styrian Salza valley in the area of the Hochschwab to Vienna. More than 100 aqueducts and 19 culverts (tube siphons functioning on the principle of communicating vessels) with length up to 2.5 km had to be built to bridge or to underpass valleys and rivers. The water takes about 36 h to reach Vienna.

2.4.1. Inauguration in 1910

On December 2, 1910, Emperor Francis Joseph I inaugurated the Second Vienna Alpine Springwater Main in a ceremonial event. On this occasion, the two fountains in the park in front of the city hall were operated with Alpine Springwater for the first time. The new main proved a blessing for the health of the population. Since the city now had fresh water in abundance, the municipality was able to set up public shower and bath houses in all districts of Vienna. In 1973 the catchment area of the springs was put under protection by declaring it a water reserve.

3. Special features of the Vienna water supply system

3.1. ALPINE SPRINGWATER

Vienna's water supply system is unique in the world. There is no other city with more than a million inhabitants which all the year round covers its water requirements almost exclusively with pure mountain springwater. The water comes directly from the Alps, more explicitly, areas in the Rax/Schneeberg and the Hochschwab region which together are twice the size of Vienna. The water is transported to Vienna through two long-distance mains with a total length of more than 300 km. The responsibility for the care and the preservation of the two water reserves is in the hands of the municipality of Vienna. The authorities take great pains to uphold the natural purification process and thus to maintain the high quality of the water. In fact, Vienna Water Works watches over the progress of the water from the moment the rain hits the ground until it arrives at its destination in a Viennese household.

3.2. WATER QUALITY

The high quality of the water the Viennese population enjoys is the result of the consistently pursued strategy of collecting top-grade water in pristine and specially protected areas and transporting it to the city. Starting with the construction of the two Alpine Springwater mains, the municipality has consistently implemented this strategy throughout the century by tapping further springs, the last one being the recently included Pfannbauern Spring.

The quality of the drinking water is monitored by the Institute of Environmental Medicine. The drinking water supplied into the distribution mains is controlled daily and the catchment areas of the First and Second Vienna Alpine Springwater Mains as well as the Third Vienna Water Main, and the Lobau groundwater works are controlled continuously (Table 11.1).

3.3. ENERGY FROM DRINKING WATER

Whereas in most countries a huge amount of energy is required to pump drinking water to the cities and, in turn, to distribute it among the city dwellers, Vienna is in the fortunate position that it obtains its water by force of gravity. The distribution of water within the city is effected by exploiting the principle of communicating vessels. Pumping stations are only needed to supply water

TABLE 11.1. Water quality of the Vienna AlpineSpringwater (2008).

Parameter	Unit	First ASWM	Second ASWM
Temperature	°C	8.0	8.4
Carbonate Hardn.	°dH	8.3	6.3
Conductibility	µS/cm	310	220
pH-Value	–	7.9	8.1
Nitrate	mg/l	5.6	3.2
Nitride	mg/l	<0.008	<0.008
AMMONIUM	mg/l	<0.02	<0.02
TOC	mg/l	0.8	0.8

Water quality of Alpine Springwater (average of 2008) (municipal department 31).

to the more elevated districts of Vienna (the so-called yellow pressure zone or pumping zone).

Map with pressure zones http://www.wien.gv.at/english/environment/watersupply/images/vienna.gif

In the headwater regions of the First and the Second Vienna Alpine Springwater Mains, Vienna Water Works operates several water power plants, which exploit the natural gradients to produce electricity. Thus, Vienna Water Works not only requires very little energy for its operations, it even produces electricity. In 2008 the company's water power plants delivered 65 million kilowatt hour of electric energy. On the other hand, total energy consumed by Vienna Water Works including all pumping stations and buildings ran to only 9 million kilowatt hour.

4. Vienna's water supply in perspective

Groundwater reserves remain important for Vienna even though there is an abundant supply of Alpine Springwater. Instances in which the city has to fall back on these reserves include times in which the outpour of the springs is low or when the Alpine Springwater mains have to be taken out of operation to carry out maintenance and repair work. They also present a valuable reserve for the case of natural disasters or large-scale pollution. Damage to technical infrastructure caused by mudflows or other major incidents may entail interruptions in the supply of springwater. In such cases, the groundwater reserves will guarantee the uninterrupted supply of drinking water. With the expansion of the groundwater works in the lower Lobau and at Moosbrunn as well as the planned construction of a new plant, Vienna Water Works aims at increasing the capacity of its groundwater works to such an extent as to allow the company to replace one of the two Alpine Springwater mains.

5. Key figures of Vienna Water Works

• Vienna Water Works supplies 1.8 million people with drinking water at an average daily consumption of 390,000 m³/day.

• Ninety seven percent of the water supplied is Alpine Springwater; the remaining 3% are made up of groundwater.

• The company operates 13 power plants, which produce about 65 million kilowatt hour of electricity.

• The water can be stored in 30 reservoirs with a total capacity of 1.65 million cubic metre.

• The total length of Vienna's supply system is 3,300 km.

• The emergency service of Vienna Water Works is available 24 h a day.

City of Vienna, MA 31 – Water Works
Grabnergasse 4-6, A-1061 Vienna
Phone: +43-1-59959-31000

or

Visit our homepage: http://www.wien.gv.at/ma31/ or
http://www.wien.gv.at/english/environment/watersupply/index.html
or send us an email: astrid.rompolt@wien.gv.at

6. Kaiserbrunn Water Supply Museum

On the occasion of the 100th anniversary of the operation of the First Vienna Alpine Springwater Main the Vienna Water Works decided to make the history and status quo of Vienna's water supply accessible to the public.

Blueprints, pictures and other exhibits allow visitors of the Kaiserbrunn Water Supply Museum to trace the history of Vienna's drinking water back to the Roman period. The main bulk of the exhibition is dedicated to the construction of the first Vienna Mountain-Water Pipeline.

Opening hours http://www.wien.gv.at/wienwasser/aktiv/kaiser.html

7. Wildalpen Water Supply Museum

The museum displays information about the history of Vienna's water supply. The focus of the exhibition lies on the construction of the Second Vienna Alpine Springwater Main. In ten exhibitions rooms the masterly technical achievements of this major construction effort are illustrated by maps, plans, pictures and original documents.

Opening hours http://www.wien.gv.at/wienwasser/wildalpen/index.html

Breinigsville, PA USA
17 December 2009
229340BV00008B/3/P